Werner Massa

Crystal Structure Determination

T0155976

Springer
Berlin
Heidelberg
New York
Hong Kong
London
Milan
Paris
Tokyo

Werner Massa

Crystal Structure Determination

Translated into English by Robert O. Gould

Second Completely Updated Edition

With 107 Figures and 13 Tables

Springer

AUTHOR

Professor Dr. Werner Massa
University of Marburg
Department of Chemistry
Hans-Meerwein-Straße
35043 Marburg
Germany
E-mail: massa@chemie.uni-marburg.de

TRANSLATOR

Dr. Robert O. Gould
University of Edinburgh
Structural Biochemistry Unit,
Michael Swann Building
Mayfield Road
Edinburgh EH9 3JR
UK
E-mail: gould@ed.ac.uk

The book was originally published in German under the title "Kristallstrukturbestimmung," 3. Auflage. © B. G. Teubner GmbH, Stuttgart/Leipzig/Wiesbaden, 2002.
1st English Edition Springer-Verlag 1999

ISBN 978-3-642-05841-7

Library of Congress Cataloging-in-Publication Data Applied For

A catalog record for this book is available from the Library of Congress. Bibliographic information published by Die Deutsche Bibliothek.
Die Deutsche Bibliothek lists this publication in Die Deutsche Nationalbibliographie; detailed bibliographic data is available in the Internet at <http://dnb.ddb.de>.

This work is subject to copyright. All rights reserved, whether the whole or part of the material is concerned, specifically the rights of translation, reprinting, reuse of illustrations, recitations, broadcasting, reproduction on microfilm or in any other way, and storage in data banks. Duplication of this publication or parts thereof is permitted only under the provisions of the German Copyright Law of September 9, 1965, in its current version, and permission for use must always be obtained from Springer-Verlag. Violations are liable for prosecution under the German Copyright Law.

Springer-Verlag Berlin Heidelberg New York
Springer-Verlag is a part of Springer Science+Business Media

springeronline.com

© Springer-Verlag Berlin Heidelberg 2010
Printed in Germany

The use of general descriptive names, registered names, trademarks, etc. in this publication does not imply, even in the absence of a specific statement, that such names are exempt from the relevant protective laws and regulations and therefore free for general use.

Cover design: Erich Kirchner, Heidelberg

Printed on acid free paper   32/3141/LT-5 4 3 2 1 0

Preface to the Second English Edition

The second English edition is largely based on the third German edition of the Teubner Studienbuch "Kristallstrukturbestimmung," which appeared in 2002. In particular, Chapter 7, dealing with experimental methods, has been extensively rewritten. In view of the huge recent advances in the use of area detector systems for single-crystal data collection, their description has replaced much of the material on "classic" methods. Similarly, the practical example (Chapter 15) now describes area-collector methods more fully. Among many other cases, the sections on Rietveld refinement, macromolecular crystallography and uses of databases have been updated. I am grateful to my colleague R. O. Gould for continuing his excellent translation of the first edition, and for the friendly and careful collaboration in achieving many large and small improvements.

Werner Massa — Marburg, November 2003

Preface to the First German Edition

Crystal structure analysis using X-rays has undergone an expansion of avalanche proportions in the last twenty years, thanks to the development of rapid and automatic means of data collection and the enormous growth of the computer hardware and software for carrying out the necessary calculations. Because of its wide applicability and its precision, it has become one of the most important tools in both organic and inorganic chemical research. Despite the fact that crystallography plays a very minor role in most undergraduate study, many students have found that in the course of graduate or even undergraduate research, they need to undertake a crystal structure determination themselves, or at least to become competent to interpret crystallographic results. Thanks to ever improving program systems, the many complex steps of a structure analysis are certainly becoming less and less difficult for the beginner to master. Nonetheless, regarding the process simply as a "black box" is fraught with danger.

This book is aimed at those students of chemistry and related subjects who wish to take a look into the black box before they step into its territory, or who simply wish to learn more of the fundamentals, the opportunities and the risks of the method. In view of the well-known fact that the likelihood a book will actually be read is inversely proportional to its number of pages, fundamentals of the method are treated here as briefly and as intuitively as possible. It seems more important that chemists should have a grasp of the basic principles and their application to a problem, than that they be in a position to understand fully the complex mathematical formalisms employed by the computer programs.

On the other hand, some aspects of the subject, which bear directly on the quality of a structure determination, are worth fuller treatment. These include discussion of a number of significant errors and the recognition and treatment of disorder and twinning. Most important crystallographic literature is available in English, but a few references in other languages, principally German, have been included.

This book is based in part on lectures and on a seminar at the University of Marburg. Consciously or unconsciously, many colleagues have made their contributions. I am particularly grateful to Professor D. Babel for many helpful suggestions and a critical reading of the manuscript. I thank Dr. K. Harms for proofreading the manuscript and Mr. C. Frommen for considerable assistance with the production of camera-ready copy using the LaTeX program. Finally, I acknowledge the help of my wife Hedwig and my children for all their assistance and patience during the preparation of this book.

Werner Massa — Marburg, April 1994

Contents

Commonly used Symbols

a, b, c	lattice constants *or* symbols for axial glide planes
a^*, b^*, c^*	reciprocal lattice constants
Å	Ångström unit ($\equiv 10^{-10}$ m)
B	Debye-Waller Factor ($\equiv 8\pi^2 U$)
d	lattice-plane spacing *or* symbol for diamond glide plane
d^*	scattering vector in reciprocal space
E	normalized structure factor
f	atomic scattering factor (formfactor)
F_c	calculated structure factor
F_o	observed structure factor
FOM	figure of merit
hkl	Miller indices
I	reflection intensity
L	Lorentz factor
M_r	mass of a mole
n	order of diffraction *or* symbol for diagonal glide plane
p	polarization factor
R	conventional residual, calculated from F_o-data
S_H	sign of a structure factor
TDS	thermal diffuse scattering
U	atomic displacement factor ($\equiv B/8\pi^2$) *or* unitary structure factor
w	weight of a structure factor
wR	weighted residual, calculated from F_o-data
wR_2	weighted residual, calculated from F_o^2-data
x, y, z	atomic coordinates
Z	number of formula units per unit cell
α, β, γ	unit cell angles
Δ	path difference for interference *or* other differences
$\Delta f', \Delta f''$	components of anomalous scattering
ε	extinction coefficient
ϑ	scattering angle
λ	X-ray wavelength
μ	absorption coefficient *or* various angles
ϱ	density
σ	standard error
Φ	phase angle of a structure factor

CHAPTER 1

Introduction

To solve a crystal structure means to determine the precise spatial arrangements of all of the atoms in a chemical compound in the crystalline state. This knowledge gives a chemist access to a large range of information, including connectivity, conformation, and accurate bond lengths and angles. In addition, it implies the stoichiometry, the density, the symmetry and the three dimensional packing of the atoms in the solid.

Since interatomic distances are in the region of 100–300 pm or 1–3 Å,[1] microscopy using visible light (wavelength λ ca. 300–700 nm) is not applicable (Fig. 1.1). In 1912, Max von Laue showed that crystals are based on a three dimensional lattice which scatters radiation with a wavelength in the vicinity of interatomic distances, i. e. X-rays with $\lambda = 50$–300 pm. The process by which this radiation, without changing its wavelength, is converted through interference by the lattice to a vast number of observable "reflections" with characteristic directions in space is called *X-ray diffraction.* The method by which the directions and the intensities of these reflections are measured, and the ordering of the atoms in the crystal deduced from them, is called *X-ray structure analysis.* The following chapter deals with the lattice properties of crystals, the starting point for the explanation of these interference phenomena.

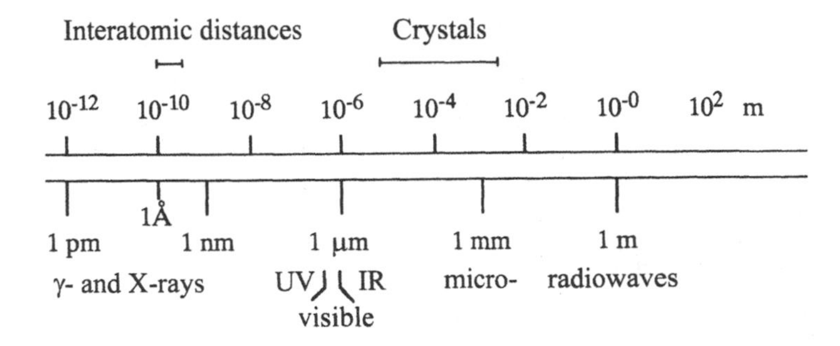

Fig. 1.1. Crystal dimensions and comparison with the wavelengths of the electromagnetic spectrum.

[1] Although not strictly S.I., the Ångström (Å) unit ≡ 100 pm, is widely used, and is almost universal in crystallographic programs.

Introduction

To solve a crystal structure means to determine the precise spatial arrangements of all of the atoms in a chemical compound in the crystalline state. This knowledge gives a chemist access to a large range of information, including connectivity, conformation, and accurate bond lengths and angles. In addition, it implies the stoichiometry, the density, the symmetry and the three dimensional packing of the atoms in the solid.

Since interatomic distances are in the region of 100–300 pm or 1–3 Å[1], microscopy using visible light (wavelength λ ca. 300–700 nm) is not applicable (Fig. 1.1). In 1912, Max von Laue showed that crystals are based on a three dimensional lattice which scatters radiation with a wavelength in the vicinity of interatomic distances, i.e. X-rays with λ = 50–300 pm. The process by which this radiation, without changing its wavelength, is converted through interference by the lattice to a vast number of observable "reflections" with characteristic directions in space is called *X-ray diffraction*. The method by which the directions and the intensities of these reflections are measured, and the ordering of the atoms in the crystal deduced from them, is called *X-ray structure analysis*. The following chapter deals with the lattice properties of crystals, the starting point for the appearance of the interference phenomenon.

Fig. 1.1. Crystals [illegible] the [illegible] elements of the electromagnetic spectrum

[1] Å is not an SI unit but the Ångström unit is still widely used and is almost universal in crystallographic programs.

CHAPTER 2

Crystal Lattices

2.1 The Lattice

A "crystal" is a solid object in which a basic pattern of atoms is repeated over and over in all three dimensions. In order to describe the structure of a crystal, it is thus only necessary to know the simplest repeating "motif" and the lengths and directions of the three vectors which together describe its repetition in space (Fig. 2.1). The motif can be a molecule, as in Fig. 2.1, or the building block of a network structure. Normally, it consists of several such units, which may be converted into one another by symmetry operations (as in Fig. 2.2). The three vectors ***a***, ***b***, ***c*** which describe the translations of the motif in space are called the basis vectors. By their operation one

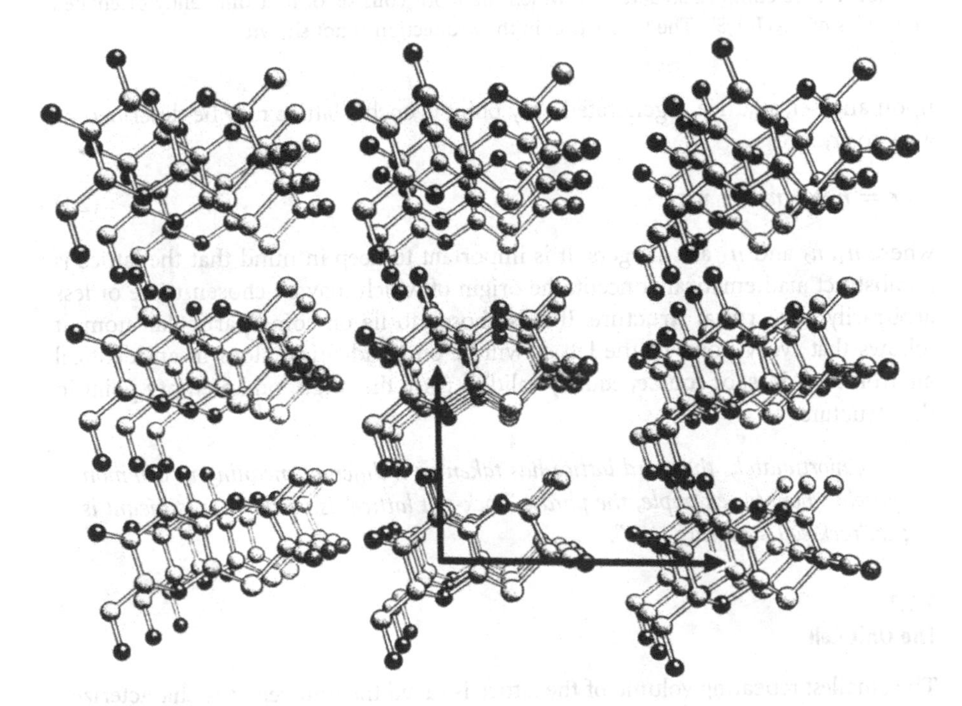

Fig. 2.1. Portion of the crystal of a simple molecular structure with the basis vectors shown. (The third vector is normal to the plane of the paper.)

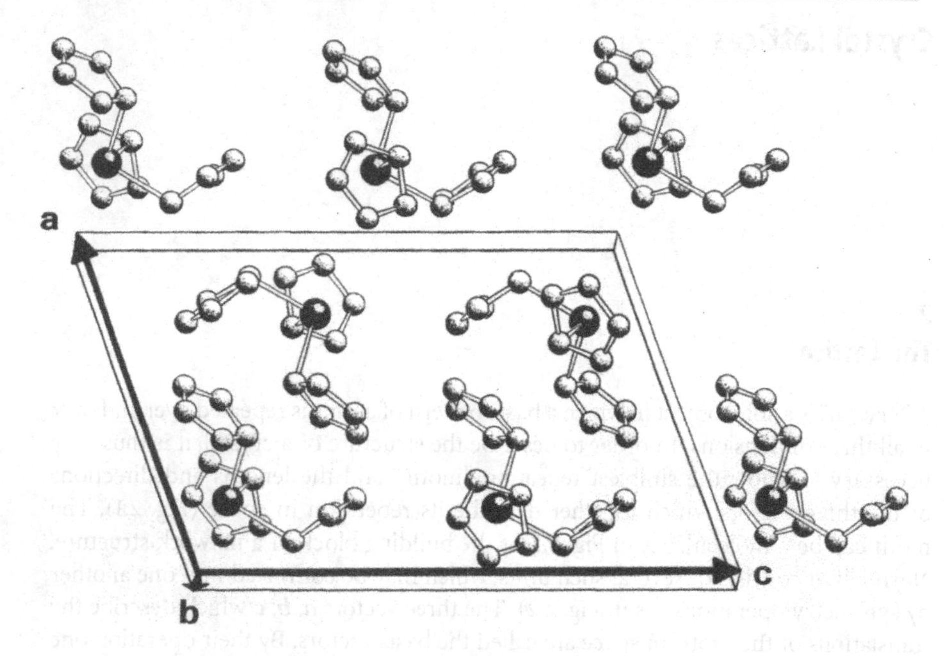

Fig. 2.2. A more complex structure in which the motif consists of four differently orientated molecules of $(C_5H_5)_3Sb$. The translation in the $\boldsymbol{b}$-direction is not shown.

upon another, a lattice is generated. Any point in such a lattice may be described by a vector $\boldsymbol{r}$,

$$\boldsymbol{r} = n_1\boldsymbol{a} + n_2\boldsymbol{b} + n_3\boldsymbol{c} \tag{2.1}$$

where n_1, n_2 and n_3 are integers. It is important to keep in mind that the lattice is an abstract mathematical concept, the origin of which may be chosen more or less arbitrarily in a crystal structure. If it is chosen to lie on some particular atom, it follows that every point of the lattice will lie on an identical atom in an identical environment. It is, of course, equally valid to place the origin on an empty point in the structure.

Unfortunately, the word lattice has taken on a different meaning in common speech: when, for example, the phrase "rock-salt lattice" is used, what is meant is the "rock-salt structure type".

2.1.1 The Unit Cell

The smallest repeating volume of the lattice is called the unit cell. It is characterized by three lattice constants a, b, c (the lengths of the basis vectors) and by the three angles α, β, γ which separate these vectors from one another. By definition, α is the angle between the basis vectors $\boldsymbol{b}$ and $\boldsymbol{c}$, β between $\boldsymbol{a}$ and $\boldsymbol{c}$, and γ between $\boldsymbol{a}$ and

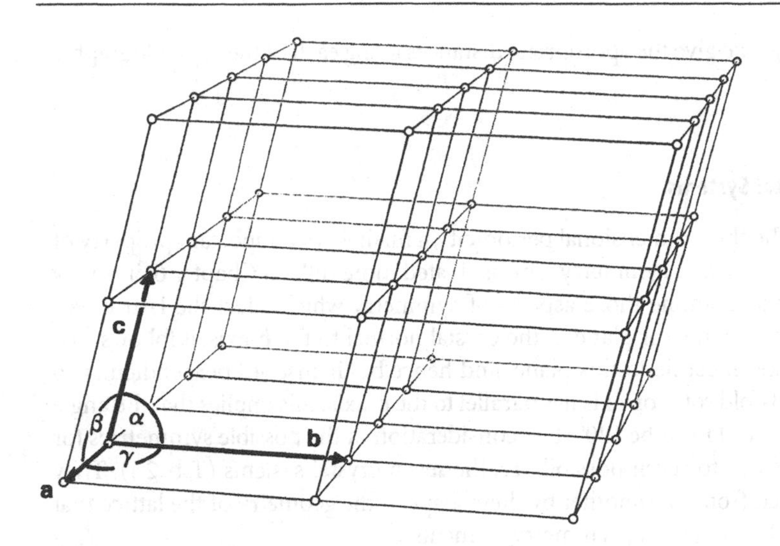

Fig. 2.3. Portion of a lattice.

b (Fig. 2.3). The lengths of the lattice constants for "normal" organic or inorganic structures, with the determination of which we are concerned here, is of the order of 3 to 40 Å. For protein structures they rise to 100 Å or more. A crystal structure is solved, if the types and locations of all the atoms in the unit cell are known; in general there will be between 1 and 1000 of these.

2.1.2 Atom Parameters

The positions of atoms are conveniently described in terms of the crystallographic axes defined by the three basis vectors: these are normally referred to as the ***a***-, ***b***- and ***c***-axes. The lattice constants are then used as units, and the atomic positions are given in terms of fractional co-ordinates *x*, *y*, *z*, which describe fractions of the lattice constants *a*, *b*, and *c* respectively (Fig. 2.4). The coordinates of an atom at the center of the unit cell, for example, are simply written as $(\frac{1}{2}, \frac{1}{2}, \frac{1}{2})$.

When a drawing is made using the published atom parameters for a structure, the lattice parameters and angles must be known. Then, "absolute" coordinates for

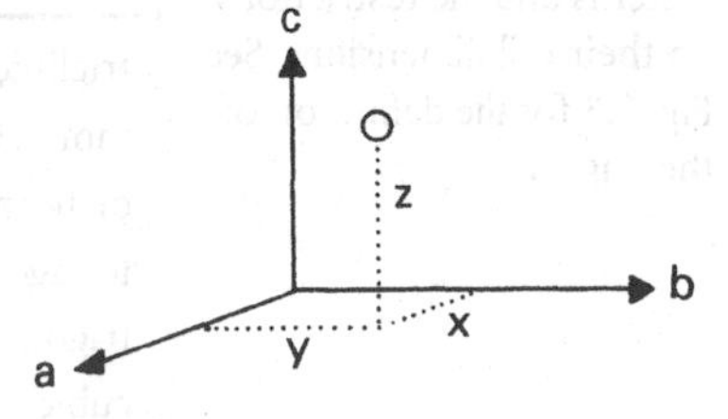

Fig. 2.4. Example of atomic parameters *x,y,z* in units of the basis vectors.

each atom xa, yb, zc give the appropriate distances along each of the crystallographic axes.

2.1.3
The Seven Crystal Systems

In addition to the three dimensional periodicity, a further very important property of nearly all crystals is their symmetry. This is treated more fully in Chapter 6; it is now only necessary to examine those aspects of symmetry which affect the lattice. For example, if there is a mirror plane in the crystal normal to the $\boldsymbol{b}$-axis, it follows that the $\boldsymbol{a}$- and $\boldsymbol{c}$-axes must lie in this plane, and hence be themselves perpendicular to the $\boldsymbol{b}$-axis. If a 3-fold rotation axis lies parallel to the $\boldsymbol{c}$-axis, this implies that the angle between $\boldsymbol{a}$ and $\boldsymbol{b}$ (γ) must be 120°. Full consideration of the possible symmetries for the lattice gives rise to seven possibilities, the seven crystal systems (Tab. 2.1). They are distinguished from one another by their shape — the geometry of the lattice that is required by the underlying symmetry elements.

Conventions: In order to describe crystal structures clearly and unambiguously, various rules have been adopted concerning the choice and naming of the unit-cell axes. In general, a "right-handed" system is chosen. This means that if the positive direction of $\boldsymbol{a}$ is directed forward and that of $\boldsymbol{b}$ to the left, then $\boldsymbol{c}$ must point upwards. If one holds the thumb, the index finger and the middle finger of the right hand as a waiter might to support a tray, then these three fingers, starting with the thumb, give the directions of the $\boldsymbol{a}$, $\boldsymbol{b}$ and $\boldsymbol{c}$-axes of a right-handed system. In the triclinic system, there are no restrictions on the choice of cell edges or angles, but in the monoclinic system, there is a "unique" axis — that one which is perpendicular to the other two. This unique axis is normally taken as the $\boldsymbol{b}$-axis, and the unrestricted angle is thus β (this is, rather inconsistently, called the second setting) and the $\boldsymbol{a}$- and $\boldsymbol{c}$-axes are chosen so that $\beta \geqslant 90°$. At one time, the $\boldsymbol{c}$-axis was chosen as the unique axis (the "first" setting — the unrestricted angle is γ). The $\boldsymbol{c}$-axis is always chosen as the unique axis in *trigonal, hexagonal* and *tetragonal* crystals.

When the unit cell of an unknown crystal is determined experimentally, its metric symmetry gives an indication of the crystal system. However, it is the actual underlying symmetry elements, which may only be fully determined at a later stage of the investigations, which determine the crystal system. That the metric symmetry

Table 2.1. The seven crystal systems and the restrictions on their cell dimensions. See Fig. 2.3 for the definition of the angles.

Restriction in	cell edges	cell angles
triclinic	none	none
monoclinic	none	$\alpha = \gamma = 90°$
orthorhombic	none	$\alpha = \beta = \gamma = 90°$
tetragonal	$a = b$	$\alpha = \beta = \gamma = 90°$
trigonal, hexagonal	$a = b$	$\alpha = \beta = 90°, \gamma = 120°$
cubic	$a = b = c$	$\alpha = \beta = \gamma = 90°$

of a crystal correspond within experimental error to the restrictions of a particular crystal system is a necessary but not a sufficient condition for establishing it. Occasionally it happens, as with the cryolites, $Na_3M^{III}F_6$, that all cell angles are within less than a tenth of a degree of 90°, but the crystal is actually not orthorhombic, but only monoclinic. The β-angle is merely very near 90° by chance.

2.2 The Fourteen Bravais Lattices

In the description of a lattice, it was said that the smallest possible basis vectors should be chosen for the crystal. The smallest possible unit in this lattice, the unit cell, is then the smallest volume that is representative of the crystal as a whole. This is called a "*primitive cell*". As is shown in Fig. 2.5, there are several ways in which this unit cell can be chosen.

All of the cells, shown here in two dimensional projection, are primitive and have the same volume. The choice of cell for the description of a crystal structure will be that by which the symmetry elements are best described. In other words, the cell which shows the highest possible symmetry. Usually, this implies the choice of orthogonal or hexagonal axial systems. The origin of the cell is located on an inversion center if that is possible. There are situations (Fig. 2.6) where all variants of a primitive unit cell are oblique, but that a larger cell, with 2, 3 or 4 times the volume, may be chosen which corresponds to a crystal system of higher symmetry. In order to be able to describe the symmetry elements conveniently, it is usually better to use the larger cells, even though they contain additional lattice points. Such cells are called centered and contain 2, 3 or 4 lattice points.

When lattices are described by these larger cells, to the six primitive lattices must be added eight centered lattices, which together are described as the fourteen Bravais lattices. Primitive lattices are given the symbol P. The symbol A is given to a one-face-centered or end-centered lattice, in which a second lattice point lies at the center of the A-face (that defined by the $\boldsymbol{b}$- and $\boldsymbol{c}$-axes), and B or C for a lattice centered

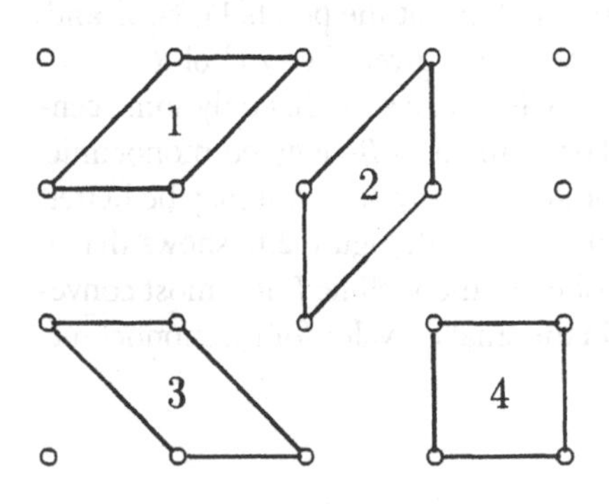

Fig. 2.5. Various choices of primitive unit cells in a lattice.

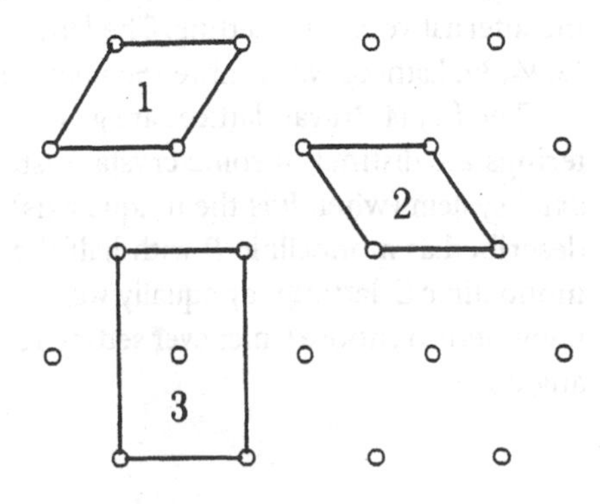

Fig. 2.6. The choice of cell 3 illustrates a centred lattice.

on the B or C face. In these cases, the cell volume is double that of the primitive cell. If the cell has lattice points at the centers of the A, B and C faces, it is called F (all face centered lattice), and has four times the volume of a primitive cell. A cell with a lattice point at its centre has double the volume of its primitive cell and is called a body centered lattice and given the symbol I (from the German innenzentriert). Nearly all metals crystallize in a cubic I or F lattice.

N.B. In the cubic CsCl structure, a unit cell may be chosen with the Cs atoms at the corners and the Cl atom at the body centre. Despite what is written in many texts, this is a primitive *cubic lattice. A body centered lattice requires that the origin and the body center of the cell be occupied by the same atoms or molecules having the same environment and the same orientation. In other words, shifting the origin of the cell to the body center must give a description of the structure indistinguishable from the original one.*

2.2.1 The Hexagonal, Trigonal and Rhombohedral Systems

Both the *hexagonal* (with 6-fold symmetry) and the trigonal (with 3-fold symmetry) systems require a hexagonal axial system, ($a = b \neq c, \alpha = \beta = 90°, \gamma = 120°$). They are conventionally described with the 6-fold axis of the lattice parallel to the $\boldsymbol{c}$-axis. For this reason, many texts recognize only six crystal systems, and treat trigonal as a subset of hexagonal. The trigonal system does, however, have one unique feature, and that is the *rhombohedral* unit cell. In this case, the smallest primitive cell may be chosen with $a = b = c, \alpha = \beta = \gamma \neq 90°$. The unique axis, along which the 3-fold symmetry axis lies, is now one of the body diagonals of the cell. In order to make this more easily described mathematically, it is convenient to transform this cell to one which is centered at the points ⅓, ⅔, ⅔ and ⅔, ⅓, ⅓, and is thus three times as large, but has the shape of the conventional hexagonal cell, with the $\boldsymbol{c}$-direction as the unique axis. (Fig. 2.7).

This is called the *obverse* setting of a rhombohedral unit cell, and is the standard setting for the rhombohedral system. Rotating the $\boldsymbol{a}$- and $\boldsymbol{b}$-axes by 60° about $\boldsymbol{c}$ gives the alternative *reverse* setting. The lattice is now centered at the points ⅓, ⅔, ⅓ and ⅔, ⅓, ⅔. Lattices which have *rhombohedral centering* are given the symbol R.

The full 14 Bravais lattices are given in Fig. 2.8. It can be seen that only some centerings are distinct in some crystal systems. For example, a B-centered monoclinic axial system (when $\boldsymbol{b}$ is the unique axis) is not given — any such cell may be better described as monoclinic P with half the volume (Fig. 2.9). Figure 2.10 shows that a monoclinic C-lattice may equally well be described as monoclinic I. It is most convenient here to choose whichever setting results in the smallest value for the monoclinic angle β.

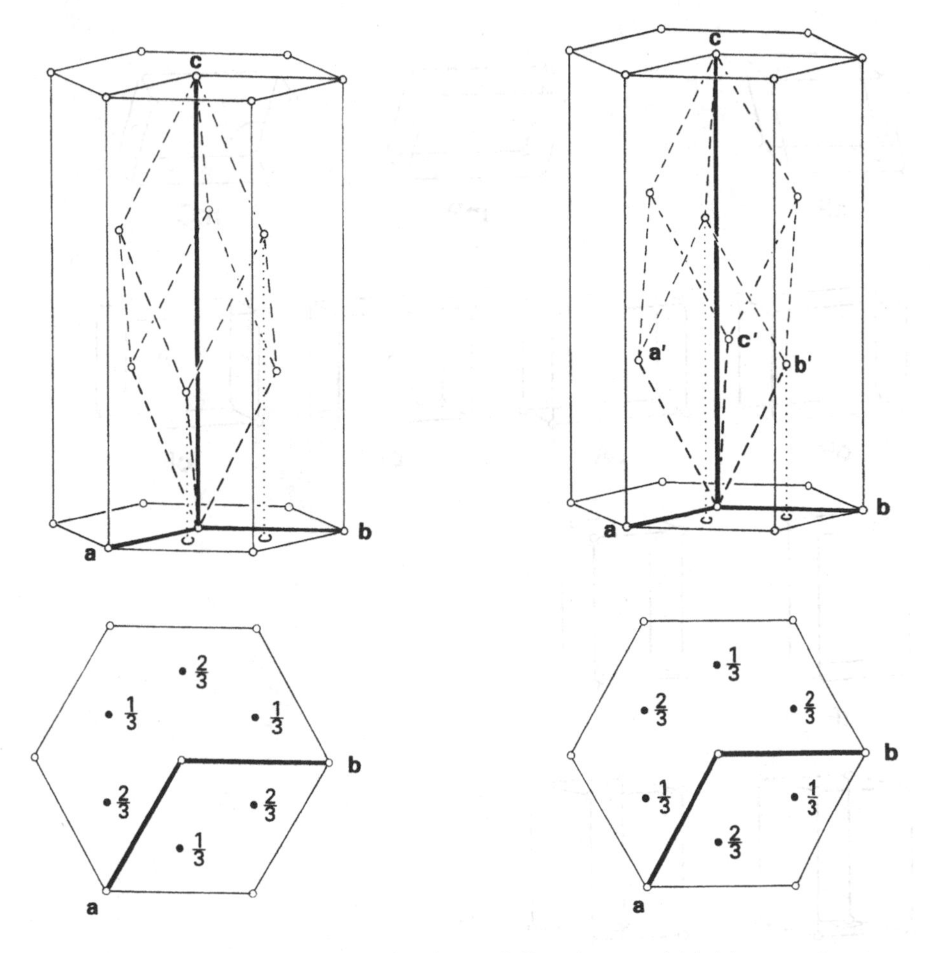

Fig. 2.7. A Rhombohedral unit cell in the *obverse* (left) and *reverse* (right) *hexagonal setting.*

2.2.2 The Reduced Cell

In order to discover whether an experimentally determined unit cell may in fact be transformed into a "better" cell of higher symmetry, algorithms have been developed to transform any cell into the so-called standard reduced form. This must fulfil the condition that $a \leqslant b \leqslant c$, and that α, β and γ are all either $\leqslant 90°$ or $\geqslant 90°$. For any crystal whatever, there is in principle only one cell which fulfils these conditions. One very important use of the reduced cell is in checking whether a particular structure has already been reported in the literature. Comparison of a reduced cell with those in data bases (see Chapter 13) should uncover any equivalent reduced cells, even if they were originally reported differently. Such a precaution should always be taken before embarking on intensity measurements (Chapter 7) for a "new" compound. A second very important use of the reduced cell is that it gives a clear guide to the

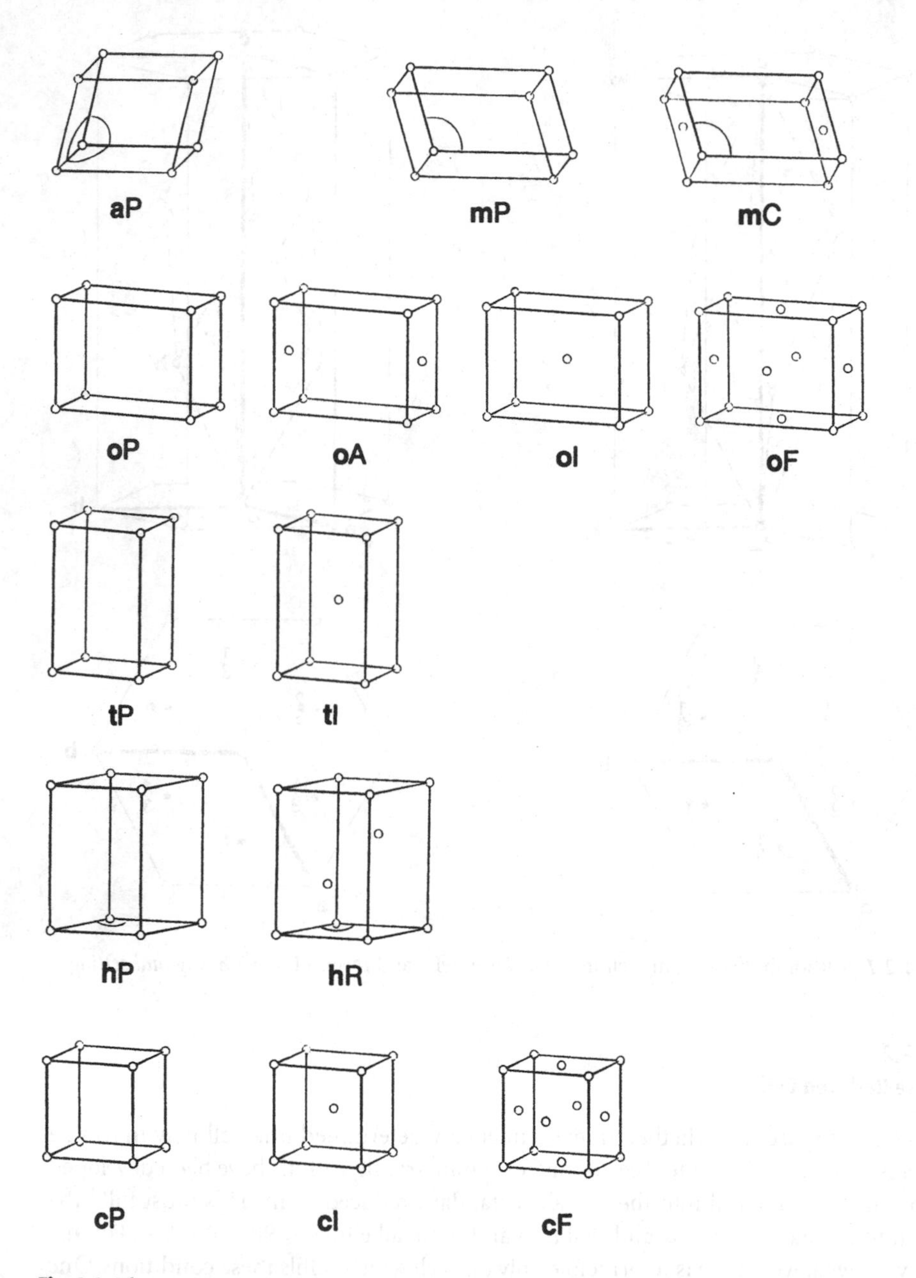

Fig. 2.8. The 14 Bravais lattices (Pearson's nomenclature). **aP** triclinic; **mP** monoclinic primitive; **mC** monoclinic C-centered (may be transformed to **mI**); **oP** orthorhombic primitive; **oA** orthorhombic A-centred (also, with different choice of axes, **oC**); **oI** orthorhombic body-centered; **oF** orthorhombic (all-)face centered; **tP** tetragonal primitive; **tI** tetragonal body-centered; **hP** trigonal or hexagonal primitive; **hR** rhombohedral, hexagonal setting; **cP** cubic primitive; **cI** cubic body-centered; **cF** cubic (all-)face centered.

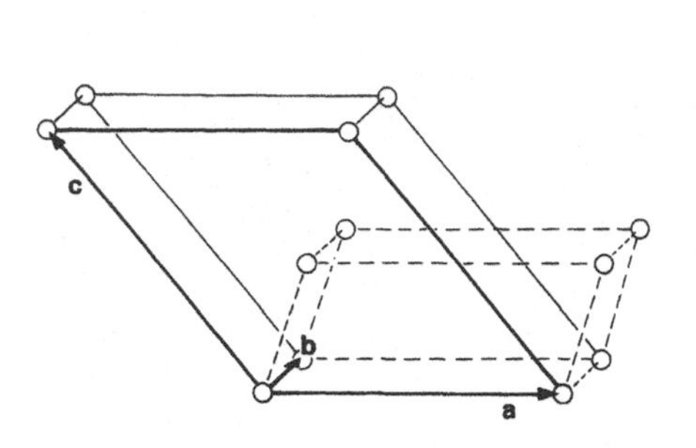

Fig. 2.9. Unnecessary monoclinic B-centering; correct P-cell outlined with dashes.

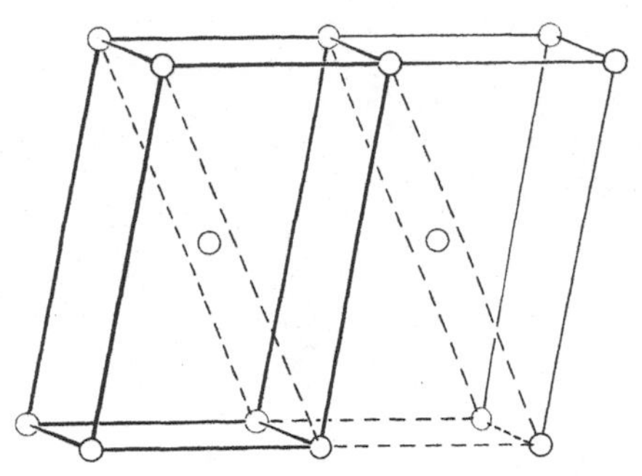

Fig. 2.10. Alternative monoclinic C- (dashed) and I-centering (full lines). In this case, I is preferred. View approximately normal to the ac-plane.

metric symmetry of the cell. This is usually expressed in terms of the Niggli-matrix (equation 2.2) which can indicate possible "correct" conventional cells. (International Tables for Crystallography, Vol. A, Chapter 9)[1] [12]

$$\text{Niggli-Matrix:} \quad \begin{pmatrix} a^2 & b^2 & c^2 \\ bc\cos\alpha & ac\cos\beta & ab\cos\gamma \end{pmatrix} \tag{2.2}$$

The reduction of a cell and its subsequent transformation to the conventional cell can be carried out by programs such as LEPAGE [56] or usually using the supplied software of a single-crystal diffractometer. It will indicate the possible Bravais lattices for a crystal. At this point, only the metric symmetry of the crystal can be established. The actual symmetry may be lower, but cannot be higher. How a unit cell is established experimentally will be discussed in chapters 3, 4 and 7.

[1] *International Tables of Crystallography* are a key resource for crystallographers. The latest edition currently consists of Volumes A, B, C, E, F (see www.iucr.org/iucr-top/it), from which Vol. A (space group symmetry) and C (mathematical, physical and chemical tables) are most important for practical work.

CHAPTER 3

The Geometry of X-Ray Diffraction

Since a crystal is a periodic, 3-dimensional array, characterised by its lattice, it should show characteristic interference phenomena. These would be expected when radiation with a wavelength of the order of the lattice spacings interacts with the crystal. In the following sections, the production of the necessary monochromatic X-rays will be described.

3.1 X-Rays

Most studies using X-rays generate them using a sealed high-vacuum tube similar to that shown in Fig. 3.1. A focused beam of electrons, generated by an applied voltage of 30–60 kV, is made to impinge on an anode (also called an "anticathode"), which is a flat plate of a very pure metal (usually Mo or Cu, less often Ag, Fe, Cr etc.) Thus, a small area (0.4 × 8 mm or 0.4 × 12 mm for "fine focus tubes" and 1 × 10 mm for "normal focus") sustains a power input of up to 3 kW, and is cooled by water. In the surface layers of the anode, X-rays are then produced by two separate mechanisms. In the first, the deceleration of the electrons by the field of the metal ions converts

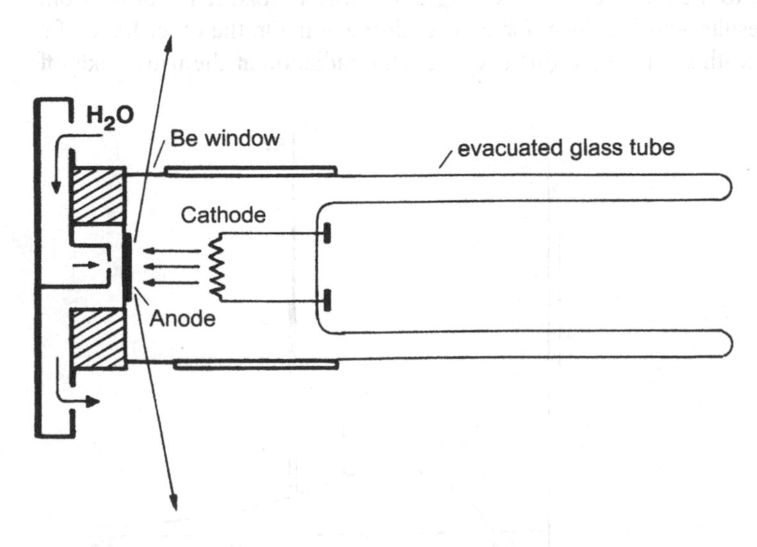

Fig. 3.1. Schematic diagram of an X-ray tube.

some of their energy into radiation. Since this gives a continuous energy spectrum, the radiation produced is called "white" radiation. The shortest wavelength that can be produced is simply related to the applied voltage:

$$\lambda_{\text{min}} = \frac{hc}{eU} \tag{3.1}$$

where h is Planck's constant, c is the velocity of light, e is the electronic charge and U is the applied voltage. If U is in kV, λ_{min} in Å is thus approximately $12.4/U$.

In addition to this white radiation, the "characteristic radiation" is produced which is much more important for the study of crystal structure. This radiation arises as a result of many electrons being expelled from atoms of the target material, in particular from the K-shell (principal quantum number $n = 1$). When an electron from a higher level (usually the L-shell, $n = 2$) falls back into the vacancy in the K-shell, an X-ray photon with a well-defined wavelength is emitted, this wavelength corresponding to the energy difference between the two levels. In terms of the angular momentum quantum number, l, and the inner quantum number j, resulting from spin-orbit coupling, the L-shell gives three possibilities, $l = 0, j = ½; l = 1, j = ½$; and $l = 1, j = 3/2$. Because of the selection rule for transitions between the K- and the L-shells ($\Delta l = \pm 1$), a closely spaced doublet is expected, known as $K_{\alpha 1}$ and $K_{\alpha 2}$ radiation. This is similar to the doublet observed for the Na-D-line in the visible region. If an electron from the M-shell falls back to the K-shell, by the same argument the higher energy doublet $K_{\beta 1}$ and $K_{\beta 2}$ is emitted. Radiation resulting from ionizations of the L- or higher shells is much weaker, and is not significant for X-ray diffraction. Fig. 3.2 shows a typical spectrum for an X-ray tube with a Mo target.

The radiation will be emitted from the line-focus in all directions, but the only useful radiation is that which leaves the tube by one of the four Be windows. If a window parallel to the line-focus is used (Fig. 3.3 down) a broad X-ray beam from the line focus results, which is ideal for powder diffraction. On the other hand, if a window at 90° to this (Fig. 3.3 right) is chosen, the radiation at the usual take-off

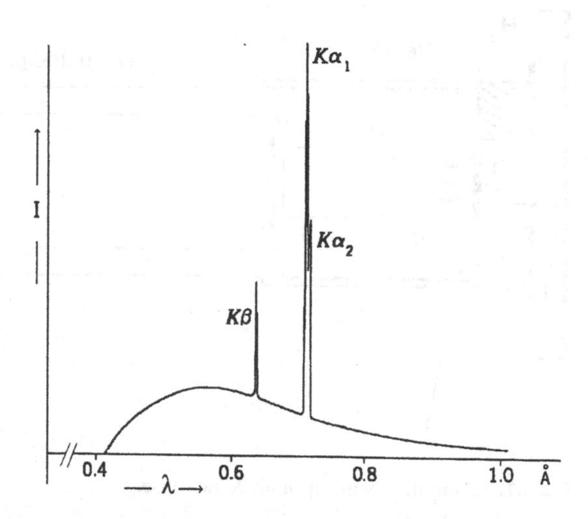

Fig. 3.2. Spectrum of a Mo-tube.

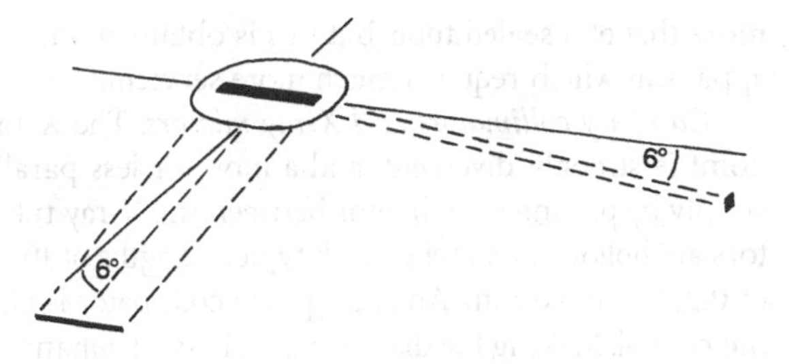

Fig. 3.3. Use of an X-ray tube as a radiation source with a line focus (down) or point focus (right).

angle of 6° to the plane of the anode will give a beam with a point focus, which is preferred as an intense radiation source for single-crystal work.

Monochromatization: Since nearly all diffraction experiments are carried out with monochromatic radiation, the very strong K_α lines are normally employed (Tab. 3.1), and it is essential to eliminate radiation of other wavelengths, particularly the K_β radiation. One way of doing this is to use a filter. These make use of the fact that metals strongly absorb X-rays when their energy is just above that required to ionise an inner electron of the metal. For example, to eliminate Cu K_β radiation but allow Cu K_α to pass, a filter of Ni foil is suitable, since the ionisation energy of the K-shell of Ni lies under the energy of Cu K_β radiation but above that of K_α. In the same way, Mn filters may be used for Fe-radiation, and Zr for Mo-radiation. By this method, relatively little of the required radiation is lost, while most of the interfering radiation is removed. A better method, for radiation of high intensity, is to use a single-crystal monochromator. This consists of a thin, single-crystal flake usually of graphite, quartz, germanium or lithium fluoride, with an area of a few cm^2, orientated to the beam so that only the desired K_α line meets the condition for constructive interference. The scattered radiation thus becomes the "primary beam" for the actual diffraction experiment.

Table 3.1. K_α-wavelengths of the most important types of X-ray tubes. (*International Tables* C, Tab. 4.2.2.1). The commonly used mean K_α-wavelength are derived from the mean of the $K_{\alpha 1}$- and $K_{\alpha 2}$-wavelengths weighted by their 2 : 1 intensity ratio.

	Mo	Cu	Fe
$K_{\alpha 1}$	0.70926	1.54051	1.93597
$K_{\alpha 2}$	0.713543	1.54433	1.93991
$K_{\overline{\alpha}}$	0.71069	1.54178	1.93728

Using bent quartz or germanium monochromators, it is even possible to separate the $K_{\alpha 1}$ and $K_{\alpha 2}$ wavelengths. For most single crystal work, this is not necessary, and in order to get the highest possible intensity, graphite monochromators are used which do not split the $K_{\alpha 1}/K_{\alpha 2}$ doublet.

Rotating anode generators. Very considerably higher intensity may be obtained by replacing the sealed high vacuum tube containing a fixed anode by an open system with a rapidly rotating anode. In this way, the heat generated is more easily carried away, and much higher power may be used. The high vacuum is obtained by continuous pumping of the system. The intensities obtained will be about three times and more that of a sealed tube, but this is obtained only with the penalty of more costly apparatus which requires much more servicing.

Capillary collimators and X-ray mirrors. The X-ray beam which leaves the focal point is strongly divergent, and a more or less parallel beam is normally obtained simply by placing a collimator between the X-ray tube and the crystal. Such collimators are hollow metal tubes with typical lengths of 10–12 cm and an internal diameter of 0.3, 0.5 or 0.8 mm. An appropriate collimator is chosen, depending on the size of the crystal. Making the diameter smaller will enhance the sharpness of the diffracted beams and reduce their breadth, but will also lower their intensity. Recently, collimators have been developed, in which the inner surface is a capillary of special glass, with such a flat surface and such a precise geometry that total reflection of X-ray beams occurs, and the beam emerges with a divergence of only a few tenths of a degree. Since a greater fraction of the incident beam is thus used, gains in intensity of over 100 % are possible. With the long wavelength Cu Kα radiation, and particularly in macromolecular crystallography and powder diffraction, collimators with multilayer mirrors have been developed which achieve a three to four-fold enhancement of intensity.

A notable advance in generator design combines such a reflecting collimator in the form of a semiellipsoid of rotation with an electronically controlled focus inside the X-ray tube. The brilliance of the resulting X-rays is such that a generator power of only 80 W can achieve the same intensity as a normal rotating anode generator consuming over 100 times the power. At present, however, there are still problems in achieving long-term stability for such systems.

Synchrotron Radiation. Instead of using the characteristic X-radiation produced by X-ray tubes, it is possible to make use of the radiation produced as a by-product of particle acceleration in a synchrotron. There are several advantages:

- very high intensity and very low divergence
- tuneable wavelengths (cf. Section 10.2)
- higher degree of polarisation.

Synchrotron sources are widely distributed around the world. The most important are probably those at the Brookhaven National Laboratory in the United States, ESRF in Grenoble, France, DESY in Hamburg, and the "Photon Factory" in Japan. The main applications of these to structure determination are for macromolecules (mainly proteins), for very small crystals, for high-resolution powder diffractometry, and for other special measurements, making use of the highly polarized beam. For further information, see reference 27.

3.2 Interference by a One-Dimensional Lattice

As a simple model for the interaction between X-rays and crystals, it is best to begin with a one-dimensional case, which can easily be modelled with an optical grating. When a ray of light with wavelength λ impinges on a grating with a spacing of d,

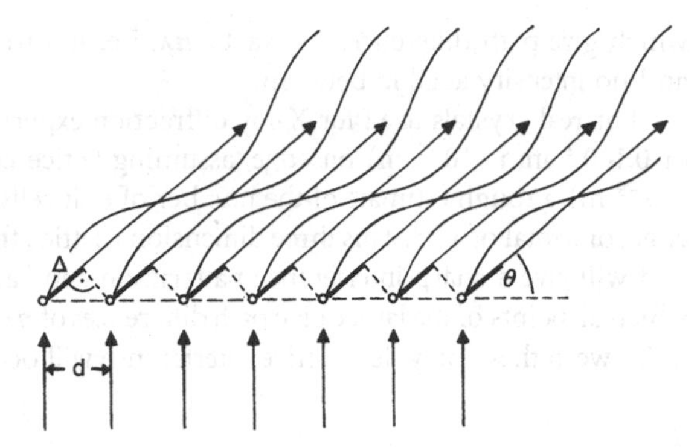

Fig. 3.4. Path difference Δ in scattering by a one-dimensional lattice.

an *interference* or *diffraction* phenomenon is observed. This can be explained by assuming that each point of the lattice scatters the radiation elastically, i.e. it emits a spherical wave of radiation of unchanged wavelength. Depending on the scattering angle ϑ and the spacing of the points d, there will be a path-difference Δ between neighboring waves (Fig. 3.4). When the angle ϑ is chosen so that this path-difference is an integral number of wavelengths ($n\lambda$) the result is positive or constructive interference; the waves diffracted from each point will be in phase with one another, and will produce a measurable diffracted ray. In this case, n is called the *order* of diffraction. Equally obvious is the situation where ϑ is chosen such that the path difference is $n\lambda + \lambda/2$. In this case, each wave will be out of phase with its neighbor, and will undergo *destructive interference.* What happens when ϑ lies somewhere between these two extreme values is most easily shown by a specific example, say the case in which neighboring points scatter waves with a path difference of $\lambda/10$ (Fig. 3.5).

With the points numbered as shown in the figure, the addition of the scattering from points 1 and 2 will give a resultant wave that is only slightly weaker than that from points with no path difference. As points further from this are considered, however, the path difference from that of point 1 increases until at point 6 it reaches a value of $\lambda/2$. Waves 1 and 6 thus cancel each other. The same cancellation occurs for the pairs 2 and 7, 3 and 8, 4 and 9 etc., with the result that no intensity is observed for this scattering angle, and destructive interference has taken place. Clearly, the number of points in the array is important. Ten points caused destructive interference to occur with a path difference of $\lambda/10$; for the same to occur much closer to the point of constructive interference, say with a path difference of only $\lambda/100$, would require 100 points. Looked at the other way round, a lattice with a very large number of points will give constructive interference only at those sharply defined scattering angles ϑ

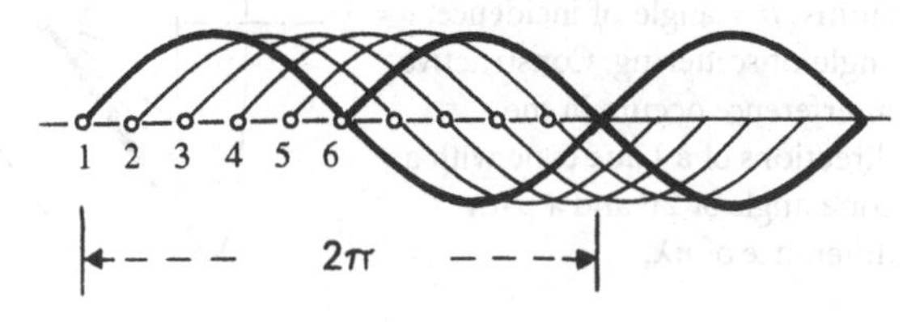

Fig. 3.5. Superposition of the scattered waves from a set with a path difference between neighbors of $\lambda/10$.

which give path differences of exactly $n\lambda$, i.e. an order of diffraction 0, 1, 2, 3 etc., and no intensity at all in between.

For real crystals used for X-ray diffraction experiments, which are of the order of 0.1–0.5 mm (10^{-4} m) on edge, assuming lattice constants of the order of 10 Å (10^{-9} m), a rough estimate of the number of unit cells would thus be 10^5 along each edge, or a total of 10^{15}. This three dimensional lattice thus contains about 10^{15} points, and will give a sharp interference pattern: only in "allowed" positions in space for which all points of the lattice give path differences of $n\lambda$ will sharp "reflections" occur; in between them only destructive interference will occur.

3.3 The Laue Equations

Before considering the results of diffraction by complex crystals, it is useful first to consider a structure as consisting of collection of "single atom" structures. In fact every atom in any crystal has the three-dimensional arrangement of the lattice points of the crystal, and the actual crystal is built up by placing together as many identical lattices as there are atoms in the unit cell, all displaced from one another as the individual atoms are. A crystal consisting of a single atom in the unit cell will thus consist of a lattice with a single scattering center at each lattice point. From such a lattice, Fig. 3.6 shows the single row of points along the a-axis. The path difference between waves scattered by two neighboring points will be related to the angle of incidence μ and the angle of scattering ν by the equation:

$$a \cos \mu_a + a \cos \nu_a = n_1 \lambda \tag{3.2}$$

For any given angle of incidence μ and any given order n_1, there will be a precisely defined scattering angle ν at which a scattered beam can be observed. Since waves are scattered in all directions, the locus of these observable waves will be a cone about the row of points with a half angle ν. For each value of n there will be one such cone, and this coaxial set are called "*Laue cones*". If the interaction of the same ray with a

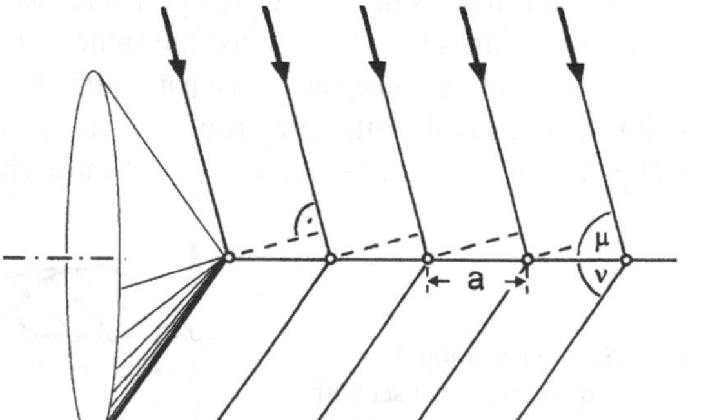

Fig. 3.6. Scattering by a row of atoms: μ = angle of incidence, ν = angle of scattering. Constructive interference occurs in the directions of a Laue cone with a cone angle of 2ν and a path difference of $n\lambda$.

second row of atoms, not parallel to the first — say, along the b-axis — is considered, the same reasoning as above will lead to equation 3.3:

$$b \cos \mu_b + b \cos \nu_b = n_2 \lambda \tag{3.3}$$

This second row of atoms will then give rise to a second set of coaxial cones indicating the directions in which observable waves are scattered. From what has already been said, it is clear that the only "allowed" directions will be those in which both conditions are met, namely those which correspond to lines of intersection of the two systems of cones.

If any specific order of diffraction n_1 from row 1 (the a-axis) is chosen, the successive points numbered along the axis from any selected origin will give the scattered rays the path differences shown in Table 3.2 and Fig. 3.7, and the same applies to the order of diffraction n_2 for row 2 (the b-axis). Choosing a point P in row 1 with the number x and a point Q in row 2 with the number y such that $x = n_2$ and $y = n_1$, the path differences will be equal (for P: $n_2 \cdot n_1\lambda$, while for Q: $n_1 \cdot n_2\lambda$). Thus, the portions of the entering and leaving wavefronts which pass through the points P and Q combine in such a way as to make the path differences for the entering and leaving ray have the same value. The scattering phenomenon thus may be described as a reflection of the ray falling on the line connecting P and Q and the angle of incidence equals the angle of reflection.

Table 3.2. Path differences in Fig. 3.7.

row 1		row 2	
atom	Δ	atom	Δ
1	$n_1\lambda$	1	$n_2\lambda$
2	$2n_1\lambda$	2	$2n_2\lambda$
3	$3n_1\lambda$	3	$3n_2\lambda$
...	...	...	...
x	$xn_1\lambda$	y	$yn_2\lambda$

In the three-dimensional case, it follows that three *Laue equations* (Equation 3.4) must apply at the same time for the direction of an incident and a diffracted ray:

$$\begin{aligned} a \cos \mu_a + a \cos \nu_a &= n_1 \lambda \\ b \cos \mu_b + b \cos \nu_b &= n_2 \lambda \\ c \cos \mu_c + c \cos \nu_c &= n_3 \lambda \end{aligned} \tag{3.4}$$

This is a most restrictive condition, and requires that the three Laue cones must intersect one another in a single line. This condition is so improbable, that it is only met when X-rays from very specific directions fall upon the crystal. For this reason, it is necessary to move a crystal in space in a quite complex way in order to observe its diffraction pattern using the single-crystal cameras and diffractometers described in Chapter 7.

It is also possible in three dimensions to describe diffraction as a reflection by a plane defined by three points of the lattice. If this reflection is "allowed," that is if the Laue conditions are met, a "reflection" is observed.

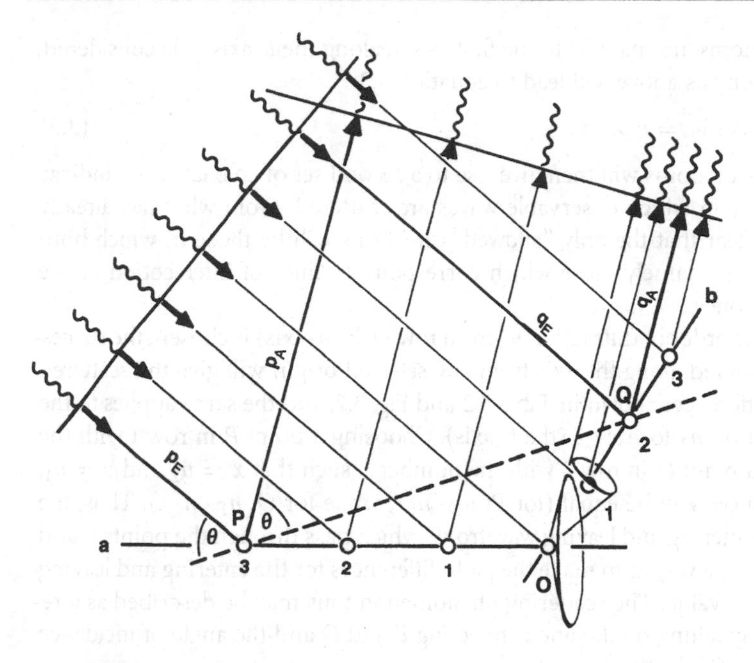

Fig. 3.7. Simultaneous fulfilment of two Laue conditions, here for second-order diffraction by the row of atoms along the a-axis and third-order diffraction by those along the b-axis. This occurs along the line of intersection of the two Laue cones, and is also shown as a reflection at the line PQ.

3.4 Lattice Planes and *hkl*-Indices

The planes which give rise to such reflections are called lattice planes, and have an orientation relative to the lattice which may be defined by *Miller indices* with the values hkl. Every plane passing through points of the lattice is one of a stack of parallel planes such that every point of the lattice will lie in one plane of the stack, regardless of its orientation. The hkl indices of any such stack can be determined by examining the plane which lies nearest to the origin, without passing through it. Its intercepts on the a-, b- and c-axes of the unit cell will be $1/h$, $1/k$ and $1/l$, which must be rational fractions (Fig. 3.8).

In the two dimensional example given in Fig. 3.9 (top right), in one set of planes, there is a plane passing through the origin and the point which is in the third row along the a-axis and the first along the b-axis; in every unit cell, these planes cut the a-axis once and the b-axis three times. Thus, considering the plane nearest the origin $^1/_h = 1/1$ and $^1/_k = 1/3$. The reciprocals of these, which are integers, are the required indices hkl. An index of 0 indicates an intercept at infinity, that is that the planes are parallel to a crystallographic axis. For example, the planes (100), (010) and (001) are parallel to the faces of the unit cell. The plane spacing d usually has its maximum value for those planes stacked along the direction of the longest lattice

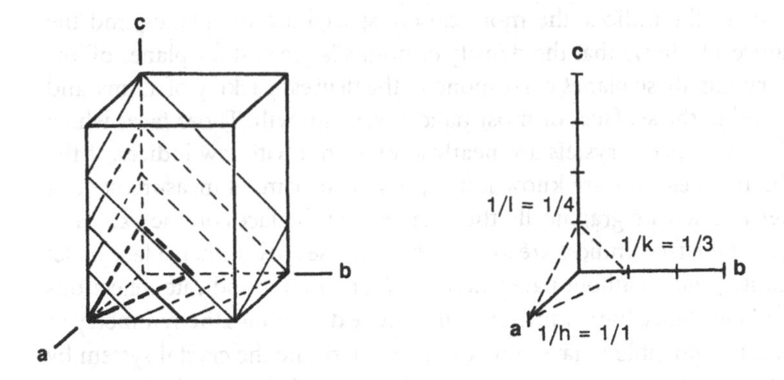

Fig. 3.8. Definition of *hkl*-values in terms of intercepts on the axes.

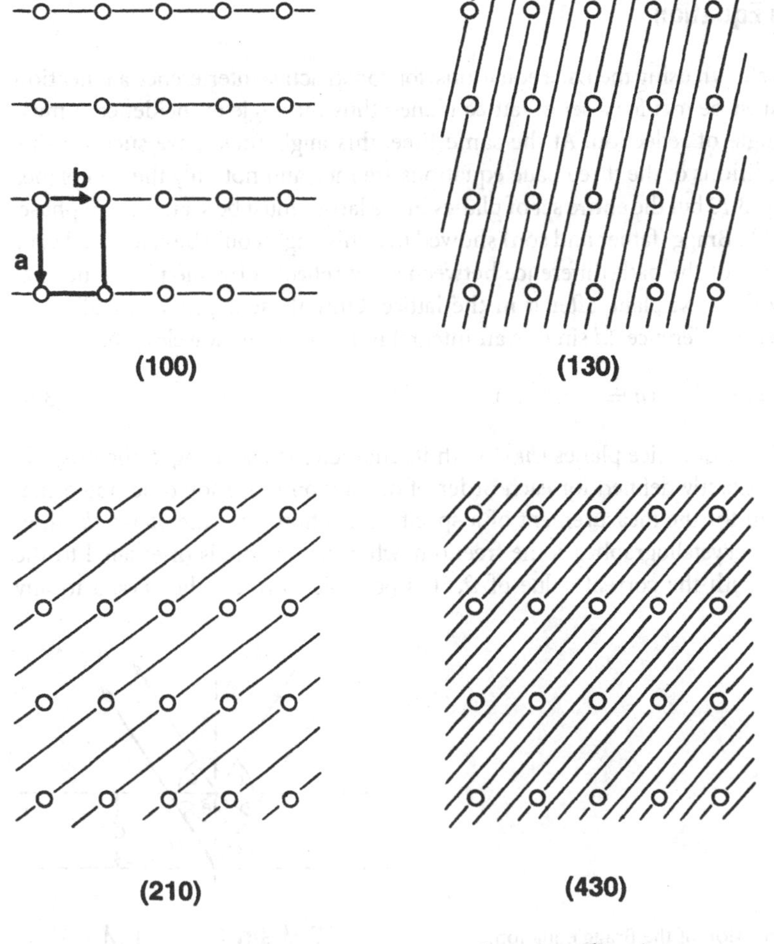

Fig. 3.9. Examples of lattice planes in projection along the *c*-direction.

constant. The higher the indices, the more closely spaced are the planes, and the smaller is d. Figure 3.9 shows that the density of points is greatest for planes of low indices. In real crystals, these planes correspond to the densest packing of atoms, and usually correspond to the surfaces of most rapid crystal growth. Those faces which are visible as the surfaces of crystals are nearly always ones with low indices. If the directions of the unit cell axes are known, it is possible by careful measurement of interfacial angles to calculate graphically the intercepts of the faces on the axes, and hence the indices of the face. Indices are given in parentheses (hkl) when a face or set of planes is meant. Indices without parentheses, hkl, are used to indicate reflections from that set of planes. Since the symmetry of the lattice determines the symmetry of the lattice planes, it is possible in favorable cases to determine the crystal system by measurement of the angles between crystal faces with a two-circle optical goniometer.

3.5 The Bragg Equation

As has been shown using the Laue equations, for constructive interference a reflection condition must be met for a set of lattice planes; thus the angle of incidence ϑ must equal the angle of reflection. At the same time, this angle must have such a value that the conditions of the three Laue equations are met, and not only the two planes shown in Fig. 3.10 but the entire set of planes in the lattice must be scattering in phase. W. H. and W. L. Bragg (father and son) showed that this angle could be calculated very easily in terms of the path difference between a ray reflected by one plane and that reflected by the next plane after it in the lattice. Only those angles ϑ are allowed, where the path difference $2d \sin \vartheta$ is an integral multiple of the wavelength:

$$2d \sin \vartheta = n\lambda \qquad (n = 1, 2, 3 \ldots) \tag{3.5}$$

For each set of lattice planes (hkl) with its characteristic spacing d, the possible angles ϑ are exactly defined for each order of diffraction n. It should be noted that this condition for the measurement of a specific reflection from the plane hkl does still leave the crystallographer some freedom: when the crystal is orientated to the X-ray beam with the correct value of ϑ, it is possible to rotate the crystal to any

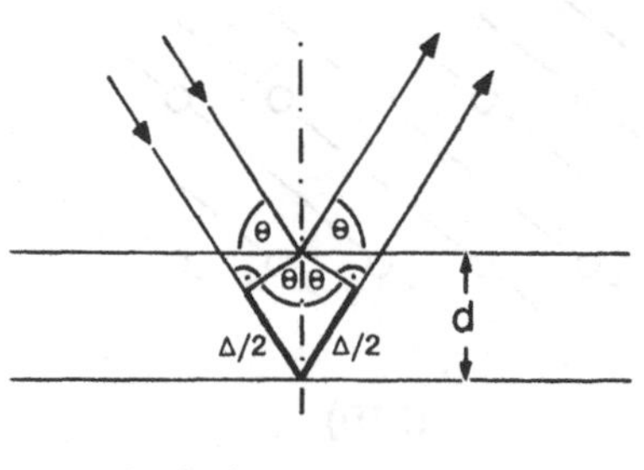

Fig. 3.10. Derivation of the Bragg equation.

position about the beam or to any position about the normal to the set of planes, as these movements affect neither the angle of incidence nor the angle of reflection.

3.6 Higher Orders of Diffraction

In order not to have to specify the order of diffraction n as well as the indices for a set of planes hkl, the concept of lattice planes has been broadened: Writing the Bragg equation (equation 3.5) in the following form:

$$2\frac{d}{n}\sin\vartheta = \lambda \qquad (n = 1, 2, 3\ldots) \tag{3.6}$$

and allowing for each "real" lattice plane hkl with spacing d further artificial lattice planes with spacings $d/2$, $d/3$, $d/4\ldots d/n$, each observable reflection can be given a unique set of indices hkl. The new artificial lattice planes naturally have new indices hkl. To describe, for example, the second order diffraction of the lattice plane (211) a corresponding lattice plane may be defined with half the spacing. This set of planes will have intercepts on the crystal axes equal to half of those of (211) and will thus have doubled indices (422). In this way, the nth order diffraction from the lattice plane hkl will be described by the artificial lattice plane $nh\ nk\ nl$ with spacing d/n. Reflections from "genuine" lattice planes have indices hkl with no common factor; if a reflection has indices with a common factor n, it gives the order of diffraction.

Maximum number of reflections. For an infinite lattice, it is obvious that the number of lattice planes will also be infinite, even without considering those of higher order. How many reflections may in practice be observed depends on the wavelength of the X-rays used. As the indices rise, the value of the spacing d falls, and the angle of diffraction given by the Bragg equation rises. The limit will be reached when that value of d is reached that corresponds to the beam being normal to the planes. This is the case when $d = \lambda/2$. Rewriting the Bragg equation in the form:

$$\sin\vartheta = \lambda/2d \tag{3.7}$$

it will be seen that this limit occurs when $\sin\vartheta$ reaches its maximum value of 1. In practice, for a crystal of average cell dimensions, several thousand reflections are measurable, and the larger the unit cell, the larger the number of accessible reflections.

3.7 The Quadratic Form of the Bragg Equation

In order to calculate the scattering angle ϑ, knowledge of the spacing of the lattice planes is necessary. This depends on the indices hkl of the lattice plane and on the geometry of the lattice. Once the unit cell of the lattice is known, the spacing d for each lattice plane hkl may be calculated from the intercepts on the axes, $^1/_h$, $^1/_k$, $^1/_l$. Since these are fractions of the lattice constants, they must first be multiplied by *a*, *b*, *c* in order to convert them to lengths. For an orthogonal, two-dimensional system (Fig. 3.11) *d* is easy to calculate: by the Law of Pythagoras, the hypotenuse of a right triangle is given by:

$$s^2 = \frac{a^2}{h^2} + \frac{b^2}{k^2} \tag{3.8}$$

The area *A* of the triangle is given by

$$2A = \frac{a}{h} \cdot \frac{b}{k} = s \cdot d \tag{3.9}$$

Squaring equation 3.9 and substituting the value for s^2 from equation 3.8 gives:

$$\frac{a^2}{h^2} \cdot \frac{b^2}{k^2} = \left(\frac{a^2}{h^2} + \frac{b^2}{k^2}\right) \cdot d^2$$

or

$$\frac{1}{d^2} = \frac{\frac{a^2}{h^2} + \frac{b^2}{k^2}}{\frac{a^2}{h^2} \cdot \frac{b^2}{k^2}} = \frac{h^2}{a^2} + \frac{k^2}{b^2} \tag{3.10}$$

In a three dimensional orthogonal (e. g. orthorhombic) system, analogous reasoning leads to equation 3.11:

$$\frac{1}{d^2} = \frac{h^2}{a^2} + \frac{k^2}{b^2} + \frac{l^2}{c^2} \tag{3.11}$$

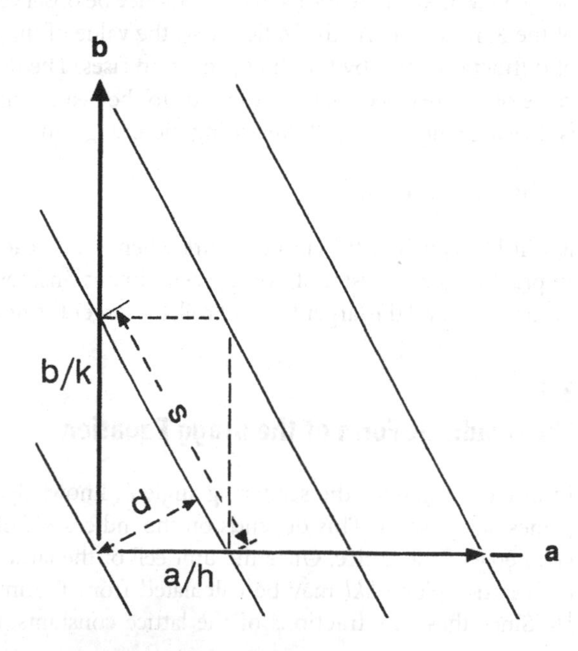

Fig. 3.11. Calculation of the interplanar spacing *d* for a set of planes *hkl* in a two-dimensional orthogonal system.

In the general, non-orthogonal cases, use must be made of the cosine law and the unit cell angles. For example, in the monoclinic system:

$$\frac{1}{d^2} = \frac{h^2}{a^2 \sin^2 \beta} + \frac{k^2}{b^2} + \frac{l^2}{c^2 \sin^2 \beta} - \frac{2hl \cos \beta}{ac \sin^2 \beta} \tag{3.12}$$

It is thus possible, knowing the unit cell, to calculate the spacing d for any set of lattice planes, and, using the Bragg equation (3.5) to calculate the scattering angle ϑ. Combination of these gives the quadratic form of the Bragg equation, which is given with the simplifications relevant to crystal systems of higher symmetry, in Table 3.3. These equations are also useful for determining lattice constants from the experimentally measured ϑ-values of selected reflections, e. g. from powder diffraction.

Table 3.3. The quadratic form of the Bragg equation in the seven crystal systems.

Triclinic

$$\sin^2 \vartheta = \frac{\lambda^2}{4}\left[h^2 a^{*2} + k^2 b^{*2} + l^2 c^{*2} + 2klb^* c^* \cos \alpha^* + 2lhc^* a^* \cos \beta^* + 2hka^* b^* \cos \gamma^*\right]$$

$$a^* = \frac{1}{V} bc \sin \alpha, \quad \cos \alpha^* = \frac{\cos \beta \cos \gamma - \cos \alpha}{\sin \beta \sin \gamma}$$

$$b^* = \frac{1}{V} ca \sin \beta, \quad \cos \beta^* = \frac{\cos \gamma \cos \alpha - \cos \beta}{\sin \gamma \sin \alpha}$$

$$c^* = \frac{1}{V} ab \sin \gamma, \quad \cos \gamma^* = \frac{\cos \alpha \cos \beta - \cos \gamma}{\sin \alpha \sin \beta}$$

$$V = abc\sqrt{1 + 2 \cos \alpha \cos \beta \cos \gamma - \cos^2 \alpha - \cos^2 \beta - \cos^2 \gamma}$$

Monoclinic

$$\sin^2 \vartheta = \frac{\lambda^2}{4}\left[\frac{h^2}{a^2 \sin^2 \beta} + \frac{k^2}{b^2} + \frac{l^2}{c^2 \sin^2 \beta} - \frac{2hl \cos \beta}{ac \sin^2 \beta}\right]$$

Orthorhombic

$$\sin^2 \vartheta = \frac{\lambda^2}{4}\left[\frac{h^2}{a^2} + \frac{k^2}{b^2} + \frac{l^2}{c^2}\right]$$

Tetragonal

$$\sin^2 \vartheta = \frac{\lambda^2}{4a^2}\left[h^2 + k^2 + \left(\frac{a}{c}\right)^2 l^2\right]$$

Hexagonal and trigonal

$$\sin^2 \vartheta = \frac{\lambda^2}{4a^2}\left[\frac{4}{3}(h^2 + k^2 + hk) + \left(\frac{a}{c}\right)^2 l^2\right]$$

Cubic

$$\sin^2 \vartheta = \frac{\lambda^2}{4a^2}\left[h^2 + k^2 + l^2\right]$$

CHAPTER 4

The Reciprocal Lattice

So far, it has been shown how the knowledge of the unit cell of a crystal makes it possible to construct all possible sets of lattice planes (*hkl*) and to calculate their spacings *d* and the scattering angle of the corresponding reflections *hkl*. In order to measure the many reflections required for a structure determination, it is necessary to determine the precise orientations of the crystal relative to the X-ray beam required to bring each set of lattice planes in turn into the position that fulfils the Bragg condition and causes the scattered radiation to fall on the detector system.

4.1 From the Direct to the Reciprocal Lattice

A drawing of the lattice of a crystal rapidly becomes incomprehensible if many sets of lattice planes are indicated on it at the same time. It would be much easier to describe each set of planes by a vector $\boldsymbol{d}$ whose direction is normal to the planes and whose length represents the plane spacing. Then each observed spot which represents a reflection on, say, a film could be identified with a point in the unit cell, specifically the end of this $\boldsymbol{d}$-vector. Because of the reciprocal relationship, $|d| \sim 1/\sin\vartheta$, the higher the *hkl* indices or the scattering angle, the shorter this vector. (For the simple example of an orthorhombic crystal, equation 3.11 applies).

$$\frac{1}{d^2} = \frac{h^2}{a^2} + \frac{k^2}{b^2} + \frac{l^2}{c^2} \tag{3.11}$$

All $\boldsymbol{d}$-vectors will thus terminate within the unit cell. Furthermore, their direction is not simple to construct, as the plane intercepts $\boldsymbol{a}/h$, $\boldsymbol{b}/k$, $\boldsymbol{c}/l$, are similarly defined in relation to the reciprocals of the indices *hkl*. This is all greatly simplified if the units $\boldsymbol{d}, \boldsymbol{a}, \boldsymbol{b}, \boldsymbol{c}$ of the direct lattice are replaced by reciprocal entities, which in *orthogonal* systems like the orthorhombic example being considered are defined simply thus:

$$d^* = \frac{1}{d}, \quad a^* = \frac{1}{a}, \quad b^* = \frac{1}{b}, \quad c^* = \frac{1}{c} \tag{4.1}$$

From this, equation 3.11 may be converted to the much simpler equation 4.2

$$d^{*2} = h^2a^{*2} + k^2b^{*2} + l^2c^{*2} \tag{4.2}$$

This has the form of a distance equation:

$$r^2 = x^2 + y^2 + z^2 \tag{4.3}$$

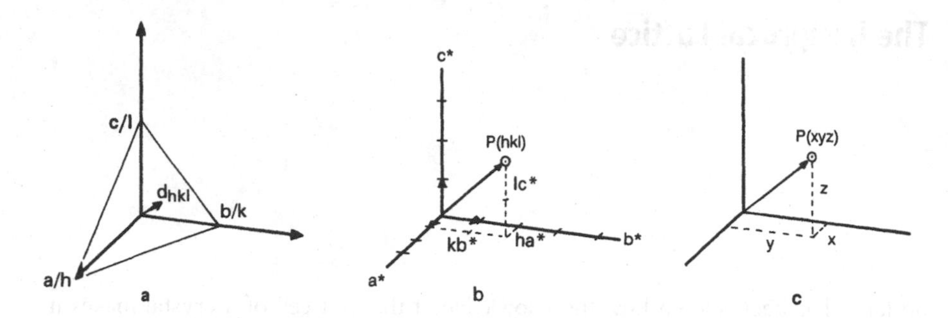

Fig. 4.1. a) $\boldsymbol{d}$-Vector in the direct lattice (scale of 10^8). b) $\boldsymbol{d}^*$-vector in the reciprocal lattice (scale $8 \cdot 10^{-8}$ cm^2) c) Comparison with a distance calculation.

Thus, in the same way that in a Cartesian coordinate system, a radius vector $\boldsymbol{r}$ defines a point with coordinates x, y, z, and has a length given by equation 4.3, a coordinate system may be defined with the reciprocal vectors $\boldsymbol{a}^*, \boldsymbol{b}^*, \boldsymbol{c}^*$ as units and the plane-spacing vector $\boldsymbol{d}^*$ easily located by its coordinates h, k, l and its length given by equation 4.2 (Fig.4.1). Since the indices *hkl* are all integral, the representation of all lattice planes by the ends of the $\boldsymbol{d}^*$ vectors gives another true lattice, resulting from three dimensional repetition of the three basis vectors $\boldsymbol{a}^*, \boldsymbol{b}^*, \boldsymbol{c}^*$. The basis vectors are now the $\boldsymbol{d}^*$ vectors for the lattice planes (100), (010) and (001). The smallest repeating three dimensional unit may thus be called the *reciprocal unit cell* and the lattice the *reciprocal lattice*. Similarly, the lattice defined by the basis vectors $\boldsymbol{a}, \boldsymbol{b}, \boldsymbol{c}$ is often called the *direct* or *real lattice*. The derivation of the reciprocal lattice is rather more complicated for oblique axial systems. This is because the normals to the (100), (010) and (001) planes and consequently the reciprocal axes $\boldsymbol{a}^*, \boldsymbol{b}^*, \boldsymbol{c}^*$ are no longer parallel to the direct axes $\boldsymbol{a}, \boldsymbol{b}, \boldsymbol{c}$. The $\boldsymbol{a}^*$-axis is thus normal to the direct $\boldsymbol{b}, \boldsymbol{c}$-plane (Fig. 4.2). This may be stated more generally: *reciprocal axes are normal to direct planes* and *direct axes are normal to the "reciprocal planes,"* defined by two reciprocal axes. Since the reciprocal axes are the normals to planes, their correct mathematical definition is given in equation 4.4 as the vector product of the required direct axes.

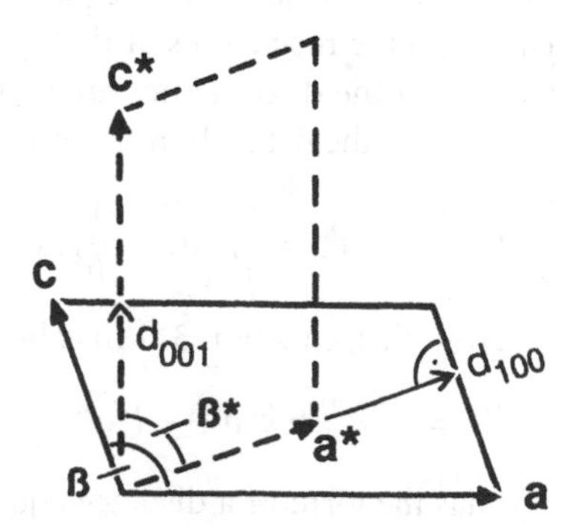

Fig. 4.2. Direct and reciprocal cell in the monoclinic crystal system.

Since these products have the dimensions of area, they are divided by the volume of the unit cell to give dimensions of reciprocal length.

$$\boldsymbol{a}^* = \frac{\boldsymbol{b} \times \boldsymbol{c}}{V}, \quad \boldsymbol{b}^* = \frac{\boldsymbol{a} \times \boldsymbol{c}}{V}, \quad \boldsymbol{c}^* = \frac{\boldsymbol{a} \times \boldsymbol{b}}{V} \tag{4.4}$$

The concept of the reciprocal lattice gives a very elegant and useful method for representing the lattice planes (*hkl*) and hence the possible reflections from a crystal (Fig. 4.3).

Every point in the reciprocal lattice represents a possible reflection *hkl*. The vector $\boldsymbol{d}^*$ is thus called the scattering vector. A diagram showing the reciprocal lattice with each point weighted to represent the intensity of the corresponding reflection, is called a(n) *(intensity)-weighted reciprocal lattice* and represents the complete "diffraction pattern" of the crystal. It is useful, particularly when considering the collection of these reflections by various techniques (Chapter 7) to consider this construction and how it may be split up into various "reciprocal planes" and "reciprocal rows". Those reciprocal planes which lie parallel to two reciprocal axes have one index constant. For example, the $\boldsymbol{a}^*$, $\boldsymbol{b}^*$-plane is called the *hk*0-plane or the 0-layer in the $\boldsymbol{c}^*$-direction. Parallel to this is the 1st layer or *hk*1-plane. Those reciprocal rows which pass through the origin, e. g. the *h*00- or *hh*0-rows (meaning $h = k$) contain the reciprocal lattice points which represent the various orders of diffraction from a "true" set of lattice planes.

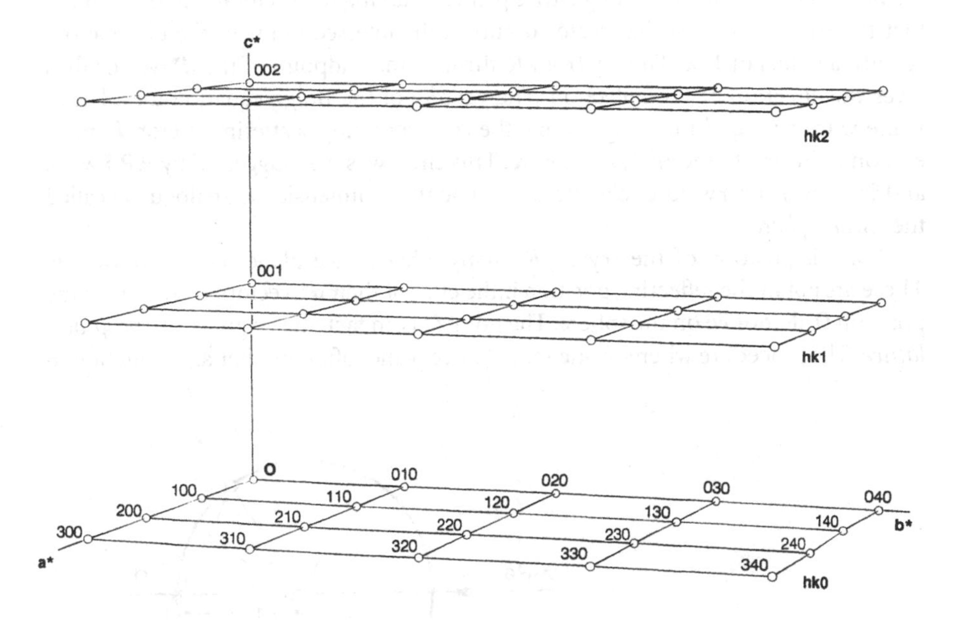

Fig. 4.3. Example of a reciprocal lattice, divided into layers along $\boldsymbol{c}^*$.

4.2 The Ewald Construction

The reciprocal lattice is not only a useful construction for displaying the lattice planes of a crystal. It is particularly well designed for the description of the practical execution of a diffraction experiment. This becomes clear from consideration of the so-called *Ewald construction* (Fig. 4.4).

On the left hand side is drawn a set of planes in the *direct* lattice. If the angle ϑ equals that required by the Bragg equation for $\boldsymbol{d}$, a diffracted beam can be observed at the angle 2ϑ. On the right side of the figure, the same condition is shown in terms of the *reciprocal* vector $\boldsymbol{d}^*$. The Bragg equation may be written in the following form:

$$\sin\vartheta = \frac{d^*/2}{1/\lambda} \tag{4.5}$$

in which the angle ϑ at the crystal $\boldsymbol{K}$ is part of a right triangle illustrating the relationship $\sin\vartheta =$ *opposite/hypotenuse*. To make this construction, a suitable scale factor must be chosen with the dimension of area. On it, in the direction of the primary beam, a line with the length of $1/\lambda$ in the direction of the primary beam is drawn from $\boldsymbol{K}$. With this as hypotenuse, a right triangle is then constructed with $\boldsymbol{d}^*/2$ as the side opposite $\boldsymbol{K}$. The adjacent side of this triangle will then fall on top of the lattice plane of the direct diagram. This is, of course, necessary, as the vector $\boldsymbol{d}^*$ is defined as the normal to the corresponding lattice plane. Doubling the vector $\boldsymbol{d}^*/2$ to $\boldsymbol{d}^*$, shows that the reciprocal scattering vector occurs at the intersection with the circle about $\boldsymbol{K}$ with a radius of $1/\lambda$. The ray from $\boldsymbol{K}$ through the endpoint of the $\boldsymbol{d}^*$-vector thus gives the direction of the reflection. The validity of the Bragg equation for a lattice plane with spacing d thus implies that the corresponding scattering vector $\boldsymbol{d}^*$ must end on a circle with radius $1/\lambda$ about $\boldsymbol{K}$. This circle was first suggested by P. P. Ewald, and is named the Ewald circle after him. The three dimensional analogue is called the *Ewald sphere.*

For this position of the crystal $\boldsymbol{K}$, many other lattice planes can be drawn in. These are not in the reflecting position if the ends of their $\boldsymbol{d}^*$ vectors, drawn from the position $\boldsymbol{O}$, do not lie on the sphere. The point $\boldsymbol{O}$ is, in fact, the origin of the *reciprocal lattice.* The procedure whereby one set of lattice planes after another are brought into

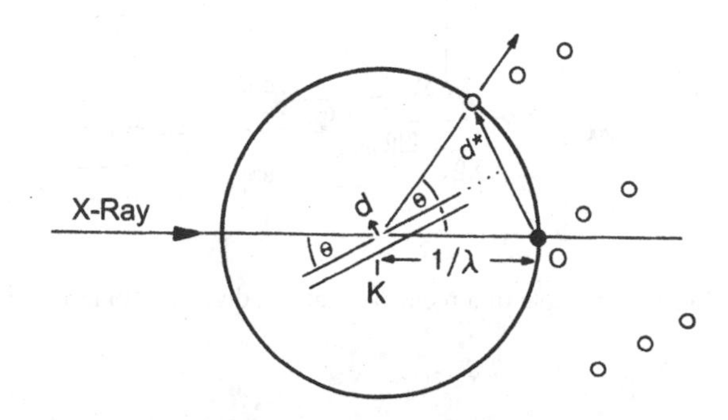

Fig. 4.4. The Ewald construction.

the reflecting position can be described in reciprocal terms as the rotation of the reciprocal lattice about the point ***O*** until one scattering vector $\boldsymbol{d}^*$ after another passes through the Ewald sphere. In every such case, there is a reflection to be measured at the corresponding angle 2ϑ. During the rotation of the crystal and its reciprocal lattice, they remain strictly parallel, as all corresponding vectors $\boldsymbol{d}$ and $\boldsymbol{d}^*$ must be parallel, despite the fact there is a little conceptual difficulty to overcome that the rotation points ***K*** and ***O*** are not coincident.

An alternative method for representing the diffraction phenomenon graphically substitutes a dimensionless Ewald sphere with a radius $R = 1$ for that with $R = 1/\lambda$ (Fig. 4.5). The diffraction vector $\boldsymbol{d}^*$ is now given as:

$$\sin\vartheta = \frac{d^* \cdot \lambda/2}{1} \tag{4.6}$$

when multiplied by the wavelength λ and thus dimensionless, it is said to be measured in *"reciprocal lattice units" (r.l.u.)*.

In the following chapter, the original Ewald construction (Fig. 4.4) will be used, since it treats the reciprocal lattice as a fixed property of the crystal. It also represents the effect on the diffraction geometry of changing the radiation wavelength – altering the radius of the Ewald sphere – more clearly.

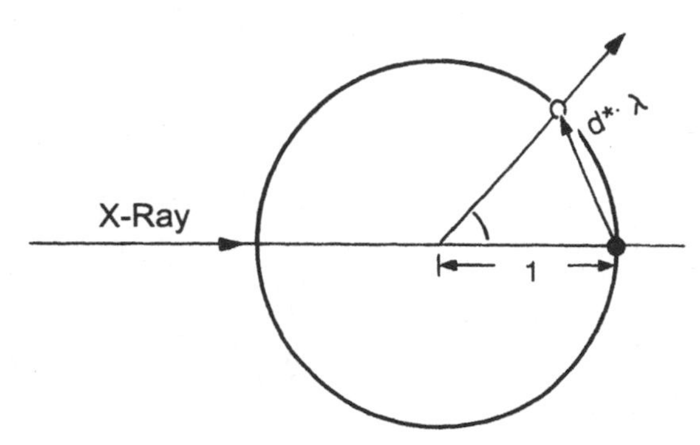

Fig. 4.5. Representation of diffraction using the dimensionless Ewald construction.

the reflecting position can be described in reciprocal terms as the rotation of the reciprocal lattice about the point O until the scattering vector d^* after another passes through the Ewald sphere. In every such case, there is a reflection to be measured at the corresponding angle 2θ. During the rotation of the crystal and its reciprocal lattice, they remain strictly parallel, as all corresponding vectors d and d^* must be parallel, despite the fact there is a little conceptual difficulty to overcome that the rotation centres C and O are not coincident.

An alternative method for representing the diffraction phenomenon graphically substitutes a dimensionless Ewald sphere with a radius of 1 for that with $R = 1/\lambda$ (Fig. 4.5). The diffraction condition is now given as

$$\sin\theta = \frac{d^*\lambda/2}{1} \qquad (4.6)$$

which is multiplied by the wavelength λ and thus dimensionless, it is said to be measured in "reciprocal lattice units" (r.l.u.).

In the following chapter, the original Ewald construction (Fig. 4.4) will be used, since it treats the reciprocal lattice as a fixed property of the crystal and represents the effect on the diffraction geometry of changing the radiation wavelength – altering the radius of the Ewald sphere – more clearly.

Fig. 4.5. [illegible] diffraction [illegible]

CHAPTER 5

Structure Factors

The previous chapter was concerned with the problem of the spatial distribution of reflections from a crystal. The question must now be addressed as to how the various intensities associated with the reflections *hkl* arise. It will then become clear how the measurement of the intensities of a large number of reflections can reveal the arrangement of atoms in the unit cell.

5.1 Atom Formfactors

It is once again useful to consider the simplest case, a "single atom structure" in which a single atom is associated with each lattice point of the crystal. Since the scattering is caused by the electrons surrounding the atom, the amplitude of the wave scattered by an atom is to a first approximation proportional to the atomic number. However, the radial fall-off in electron density from the nucleus of atoms is similar in magnitude to the wavelength of the radiation being used, so the picture of an atom as a point scatterer, which was used to develop the Laue and Bragg equations, requires some modification. The electron density must be divided into small volume elements, and their individual scattering power considered. This can be done by quantum mechanical calculations (mostly by the Hartree-Fock method, but for heavier elements by that of Dirac) of the radial dependence of electron density. Further, the distance of the element from the lattice planes being considered is important, as only electron density exactly on the planes will contribute strictly in phase to the scattered radiation. Consider a scattering center moving from one plane to the next in the stack; the path difference for the radiation would shift from 0 to λ, and the phase angle from 0 to 360 degrees or from 0 to 2π radians. A distance δ from ideal position would lead to a phase shift of $\delta \cdot 2\pi/d$, and this becomes greater, the smaller the plane spacing d, or in terms of the Bragg equation, the larger $\sin\vartheta/\lambda$ is (Fig. 5.1).

If the contributions to the scattering are summed over all volume elements, in terms of this phase difference, the result is the *ϑ-dependent scattering amplitude f* of the atom (Fig. 5.2). These are normalized for the number of electrons: at $\vartheta = 0$, f has the value of the atomic number. This ϑ-dependence varies according to the differences in electron density for individual atoms. Since they are thus related to the "form" of the atom, they are often called *atom formfactors*. These values have been tabulated both explicitly and in 9-parameter polynomial form for virtually all atoms and ions (*International Tables* C, Tab. 6.1.1.1–5) and they are stored in virtually

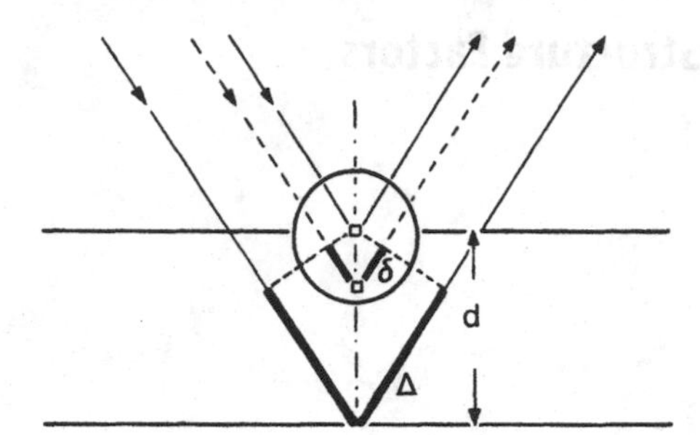

Fig. 5.1. A phase shift resulting from scattering by different points within the electron shell.

all modern programs for crystal structure analysis. These theoretically calculated quantities can be tested by experiments which generally agree with them within a few percent. These do, however, indicate the quality of *ab initio* calculations; normally the errors are greater in the experiments!

Examination of the shapes of the formfactor curves in Fig. 5.2 shows how rapidly the intensity of reflections must decrease with increasing diffraction angle. In "light atom structures" such as pure organic compounds, it is rarely useful to measure data

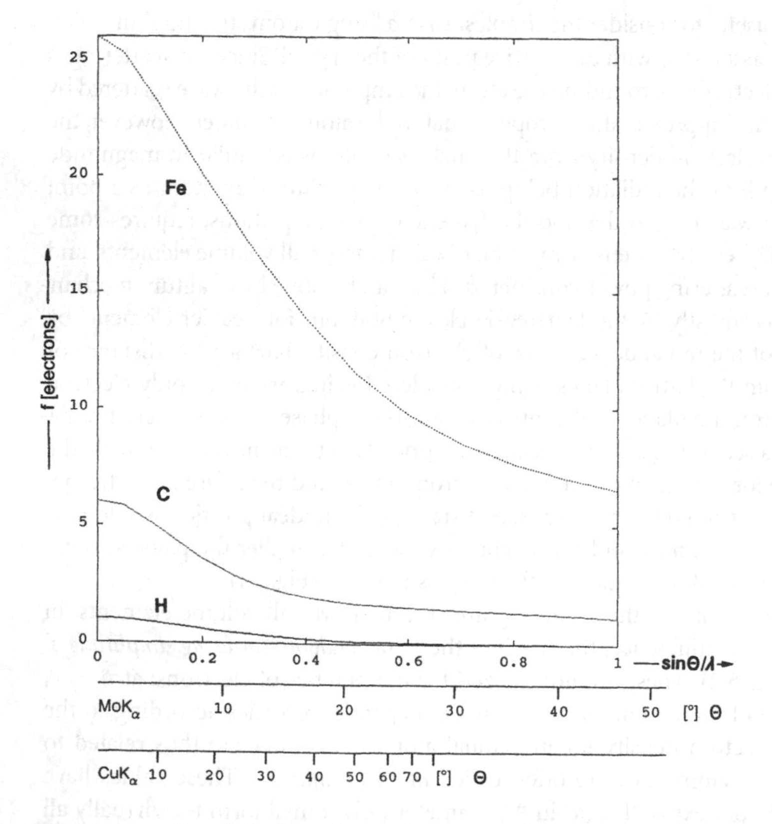

Fig. 5.2. Examples of the dependence of atom formfactors on scattering angle.

to higher ϑ-angles than 25° for Mo K_α or 70° for Cu K_α radiation. This also shows why H-atoms are so poorly located by X-ray studies. If they are near very heavy atoms, e.g. in UH_3, they are essentially invisible.

Since the core-electrons in the inner, closed shells represent a much greater electron density than do the much more spread-out *valence electrons,* the distinction between a neutral atom and its ions is very small except at the smallest scattering angles. In practice, formfactors for neutral atoms are used for nearly all studies.

5.2 Atom Displacement Factors

So far, it has been assumed that the atoms in a crystal sit in fixed positions that can be described in terms of the points of the lattice, and that they have their centers of gravity exactly on the lattice planes. In reality, the atoms are executing more or less substantial vibrations about this mean position. X-rays consist of a series of very short (ca. 10^{-18} s) bursts of radiation, since each impact of an electron on the anode of the X-ray tube releases a photon that can be treated as a coherent wavefront scarcely 100 Å or 10 nm long. This interaction time is much shorter than that of a typical thermal vibration (ca. 10^{-14} s). The final measurement is the sum (or the average) over a huge number of such single events, but each interaction is, in fact a "snapshot" of the instantaneous location of the center of the electron density.

For this reason, lattice planes are in a way "rough" (Fig. 5.3) and give rise to a situation analogous to that of the atom formfactors f (Fig. 5.1) in that the phase shift is greater the greater the magnitude of the vibration amplitude and the smaller the plane spacing d (or alternatively, the greater the scattering angle ϑ). The atom formfactors, which themselves fall off with increasing scattering angle, experience an additional weakening from atomic vibrations what can be represented by the exponential function in equation 5.1. This formula implies that the atom is "isotropic," i.e. that it vibrates equally in all directions.

$$f' = f \cdot \exp\left\{\frac{-2\pi^2 u^2}{d^2}\right\} \tag{5.1}$$

The square of the mean vibration amplitude, u^2, is called the "atomic displacement factor" U. Since it increases with temperature, it is often also called the "temperature

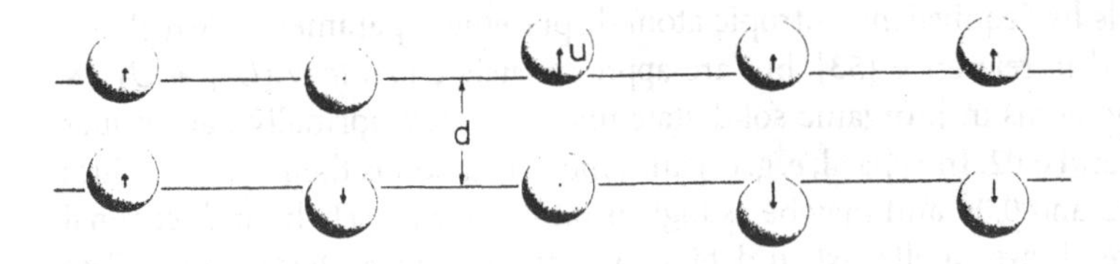

Fig. 5.3. Phase shift due to thermal motions of atoms.

factor"of the atom. Usually, d in equation 5.1 is replaced by d^* or $\sin\vartheta/\lambda$, giving the forms 5.2 and 5.3.

$$f' = f \cdot \exp\{-2\pi^2 U d^{*2}\} \tag{5.2}$$

$$f' = f \cdot \exp\left\{-8\pi^2 U \frac{\sin^2\vartheta}{\lambda^2}\right\} \tag{5.3}$$

Often, the factor $8\pi^2$ in (5.3) is also included in the isotropic displacement factor as in equation 5.4, and renamed the *"Debye-Waller factor" B*, which is significant in other fields of physics.

$$f' = f \cdot \exp\left\{-B \cdot \frac{\sin^2\vartheta}{\lambda^2}\right\} \tag{5.4}$$

Compared with isotropic vibrations, it is, not surprisingly, much more difficult to describe *anisotropic vibration*.In reality, it is normal that the amplitudes of vibrations are dependent on the direction in space. For example, the O-atom in a carbonyl group will vibrate much less strongly along the C=O bond than normal to it. The description of such a direction dependent quantity requires the use of a tensor, i. e. a quantity which is defined in magnitude and orientation with respect to three mutually perpendicular vectors. Anisotropic vibrations can be represented by so-called vibration ellipsoids with three principal axes U_1, U_2 and U_3. The shape and orientation may be given by the six U_{ij}-parameters in equation 5.5, which differs from equation 5.2 in that the scattering vector d^* has been replaced by its reciprocal lattice components (see equation 4.2 for the triclinic system)

$$f' = f \cdot e^{-2\pi^2(U_{11}h^2a^{*2} + U_{22}k^2b^{*2} + U_{33}l^2c^{*2} + 2U_{23}klb^*c^* + 2U_{13}hla^*c^* + 2U_{12}hka^*b^*)} \tag{5.5}$$

Each of the six components will have a magnitude. In an orthogonal system, the "diagonal" components U_{11}, U_{22} and U_{33} are mutually perpendicular like the principal axes of the vibration ellipsoid U_1, U_2 and U_3, and represent the extent of vibration along ***a***, ***b***, and ***c*** respectively. The "off-diagonal" terms, U_{ij} refer the orientation of the ellipsoid to that of the reciprocal axes. In non-orthogonal systems, they also have a component related to the lengths of the principal axes. The U_{ij} values are usually reported in units of Å^2 or 10^{-20} m^2. For space considerations, they are replaced in most journals by "equivalent isotropic atom displacement parameters" which are properly defined in reference [53], but are approximately equal to ⅓$[U_{11} + U_{22} + U_{33}]$. For heavy atoms in inorganic solid-state materials they normally have values between 0.005 and 0.02. In typical, e. g. organic, molecular solids their typical values lie between 0.02 and 0.06 and may be as high as 0.1–0.2 for loosely bound terminal atoms. They are always smaller when data are measured at lower temperature. The displacement parameters for all atoms are refined in the course of a structure analysis, and will be further considered in Chapter 9. In drawings of structures (using e. g.

ORTEP [58], SHELXTL [74], PLATON [70] or DIAMOND [78]), atoms are often very effectively represented as thermal ellipsoids. For the purpose, the principal axes U_1, U_2 and U_3 are scaled such that the ellipsoid represents the space in which the center of electron density can be localized with a specific probability (normally 50 %, Fig. 5.4).

Older literature and some programs use β_{ij} values in place of U_{ij}. These have absorbed not only the $2\pi^2$ factor, but also the reciprocal axes (Equation 5.6). Thus, the values are dependent on the size of the unit cell and are not easily comparable from one structure to another.

$$\beta_{11} = 2\pi^2 U_{11} a^{*2} \qquad \text{etc.} \tag{5.6}$$

By including the effect of spatial distribution of electrons about the atom in the atomic formfactor and of the thermal motions of the atoms in the displacement factors, it is now possible to calculate scattering amplitudes theoretically using the simple point description of an atom center, the positional parameters x, y, z.

5.3 Structure Factors

For a structure with only one atom in the unit cell, (Atom 1) with a position x_1, y_1, z_1 at the origin, 0, 0, 0, the scattering amplitude $F_{c(1)}$ may be calculated for each reflection *hkl*, providing the displacement parameter and the unit cell are known (the

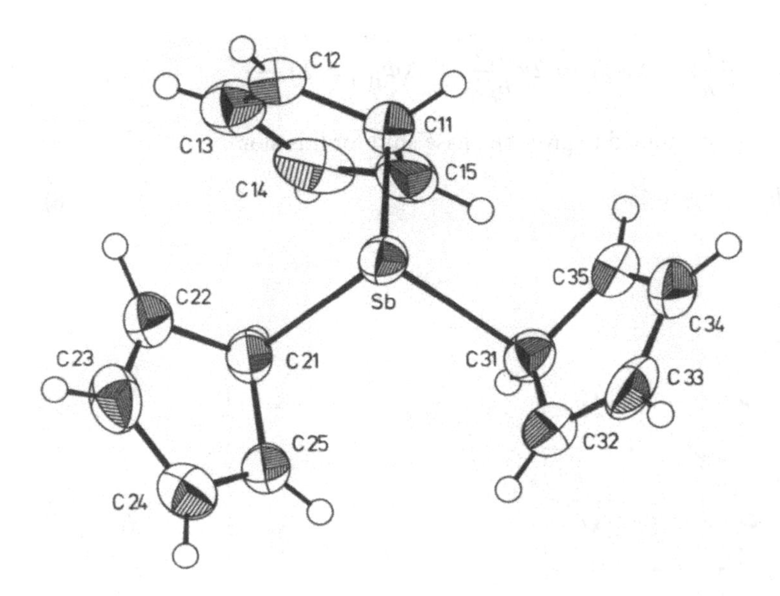

Fig. 5.4. Representation of the atoms of $(C_5H_5)_3Sb$ as 50 % probability vibration ellipsoids. Hydrogen atoms are shown as small circles of arbitrary radius.

diffraction angle for each reflection may be calculated by the formulae in Table 3.3). The formfactor f_1 for the appropriate atom type and scattering angle is taken from International Tables C, Table 6.1.1.1–5, and multiplied by the exponential function for the atomic displacement:

$$F_{c(1)} = f_1 \cdot \exp\left\{-2\pi^2 U d^{*2}\right\} \tag{5.7}$$

What is the effect of introducing a second atom type to the unit cell? For this atom, the lattice is, of course, unchanged, so the diffraction geometry is identical (Fig. 5.5). When the crystal is presented to the X-ray beam so that any particular set of planes (*hkl*) are in the diffracting position, all the atoms of type 1 will be scattering in phase with one another. The same is true of the atoms of type 2. However, because the second lattice has been shifted by a vector x_2, y_2, z_2 from the origin, the second resultant scattered wave will have a phase shift relative to the first which will be different for each reflection. By analogy, the same would be true for any further atom type i. In the same way that the scattering-angle dependence of the atomic formfactor and the displacement factor must be considered, so must the distance that separates each atom i from the lattice planes. In Fig. 5.6, the intercepts of a lattice plane *hkl* on the ***a*** and ***b*** axes are shown. Movement from one plane to the next along ***a*** or ***b*** corresponds to a phase shift moving from 0 to 2π.

The phase shift for the wave scattered by atom type i, Φ_i relative to the origin of the unit cell may be considered in three parts, and calculated by considering that the atom parameters x_i, y_i, z_i give shifts of ax_i, by_i, cz_i along the three axial directions which must then be divided by the components of the plane separation, a/h, b/k, c/l. This then gives:

$$\Delta\Phi_{i(a)} = 2\pi \frac{x_i a}{a/h}; \quad \Delta\Phi_{i(b)} = 2\pi \frac{y_i b}{b/k}; \quad \Delta\Phi_{i(c)} = 2\pi \frac{z_i c}{c/l}$$

which then may be combined to give the phase shift for this atom:

$$\Phi_i = 2\pi(hx_i + ky_i + lz_i) \tag{5.8}$$

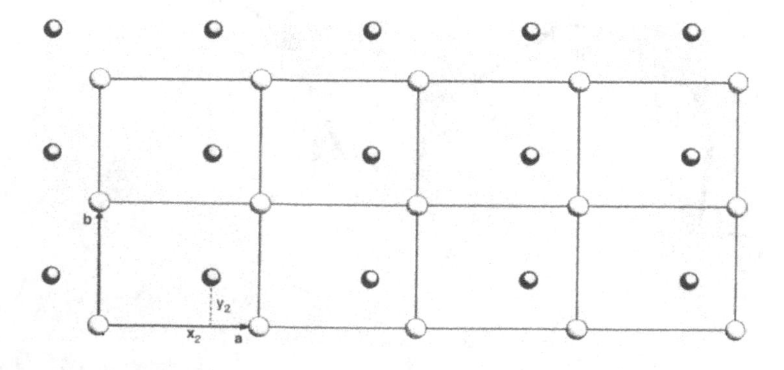

Fig. 5.5. A "two-atom structure" shown as two identical lattices displaced from one another.

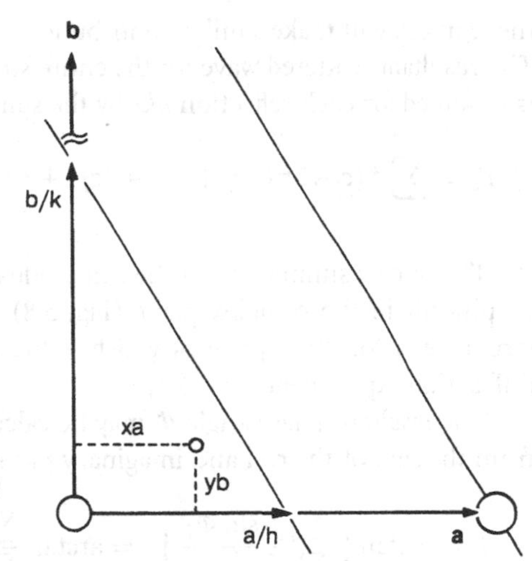

Fig. 5.6. Calculation of the phase shift of the wave scattered by atom 2 with respect to that scattered by atom 1.

Because of this phase shift, the scattered wave may be described as a *complex quantity,* and thus represented either as an exponential quantity with an imaginary exponent, (Equation 5.9) or according to the Euler formalism as the sum of a cosine term (the real part, A) and a sine term (the imaginary part, B). In the complex plane, the atom formfactor thus may be written as the sum of A and B (Fig. 5.7).

$$F_c(\text{Atom } i) = f_i \cdot e^{i\Phi_i} \tag{5.9}$$

$$F_c(\text{Atom } i) = f_i(\cos\Phi_i + i\sin\Phi_i) = A_i + iB_i \tag{5.10}$$

$$(A_i = f_i\cos\Phi_i;\; B_i = f_i\sin\Phi_i)$$

In the resulting triangle, the combination of the real and imaginary parts readily shows their relationship to both the scattering amplitude and its phase (equation 5.11):

$$|F| = \sqrt{A^2 + B^2} \qquad \Phi = \arctan\frac{B}{A} \tag{5.11}$$

This relationship applies to any particular atom type i which is separated from the origin of the unit cell by the coordinates x_i, y_i, z_i. Consequently, all of the i atoms in

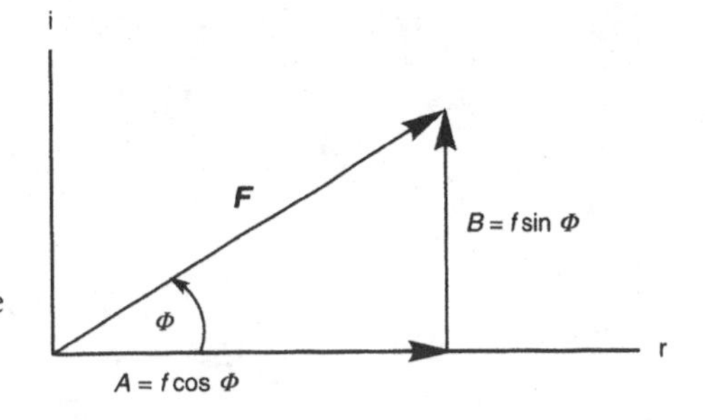

Fig. 5.7. Representation of a diffracted wave on the complex plane (r = real, i = imaginary axis).

the unit cell will make similar contributions, each with their individual phase shifts. The resultant scattered wave for the entire structure is called the *structure factor* and is obtained for each reflection *hkl* by the summation of equation 5.12:

$$F_c = \sum_i f_i \{\cos 2\pi(hx_i + ky_i + lz_i) + i \sin 2\pi(hx_i + ky_i + lz_i)\} \tag{5.12}$$

This vector summation of the *i* individual atom formfactors may be carried out graphically in the complex plane (Fig. 5.8). The result gives the amplitude of the structure factor, the square of which is the intensity which may be observed in a diffraction experiment.

The resultant phase angle Φ may be calculated by the analogue of equation 5.11 from the sum of the real and imaginary parts.

$$\Phi = \arctan\left(\frac{\sum_i f_i \sin \Phi_i}{\sum_i f_i \cos \Phi_i}\right) = \arctan \frac{\sum B_i}{\sum A_i} \tag{5.13}$$

It is, however, not directly observable experimentally, since the measured intensities only give the amplitudes of the structure factors. This constitutes the so-called *phase problem of X-ray structure analysis*, and will be discussed further under methods of structure determination in Chapter 8. Before that, however, it is essential to describe a property of crystals that is of great importance from many points of view: their symmetry.

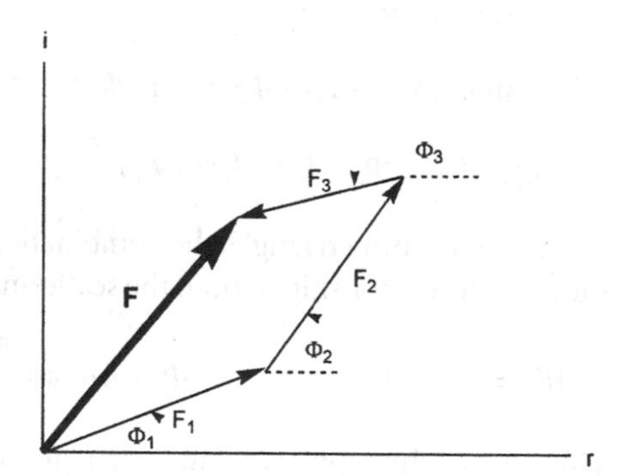

Fig. 5.8. Vectorial addition of the waves scattered by individual atoms to give the structure factor, ***F***.

CHAPTER 6

Crystal Symmetry

It is important from many points of view to understand and to be able to describe clearly the symmetry of a crystal. For example, the knowledge that there is a mirror plane in the unit cell of a structure implies that the positions of only half the atoms need be determined in order to fully describe the structure. Knowledge that there is an inversion center in the diffraction pattern makes it necessary to measure only half of the available reflections. The symmetry of a crystal has many effects on its physical characteristics, such as its optical and electrical properties. Finally, errors in determining the symmetry of a crystal can lead to its structure being impossible or difficult to solve. From almost any point of view, it is crucial to determine the symmetry of a crystal and to describe it correctly.

6.1 Simple Symmetry Elements

Symmetry elements define the (conceptual) motion of an object in space the carrying out of which, *the symmetry operation,* leads to an arrangement that is indistinguishable from the initial arrangement. The most important simple symmetry elements in crystallography are summarized in Fig. 6.1. They are referred to in crystallography by their *Hermann-Mauguin* or *International symbols,* while spectroscopists mainly use the older *Schoenflies symbols.* In International Tables A, the positions of the symmetry elements are indicated by specific graphical symbols, the most important of which are shown in Table 6.2, section 6.4.

The operation of an *inversion center* (also called a *center of symmetry*) $\bar{1}$ ("bar one" or "one bar") and the *mirror plane m* generate the mirror image of a body such as a molecule, while *rotation axes* do not. Of these, only 2-, 3-, 4- and 6-fold axes (symbols 2, 3, 4, 6) are possible in crystals, since any symmetry element must operate on the entire lattice. Regular unit cells with 5-, 7-, 8- etc. fold symmetry cannot be made to fill space. This does not mean, of course, that molecules with five-fold point symmetry cannot crystallise. It does mean, however, that such symmetry elements cannot apply to the lattice and hence the environments of these molecules will not have their molecular point symmetry. (Materials called quasicrystals have been described with these and higher order rotation axes as intrinsic symmetry elements. See section 10.1.5)

These simple operations can be *coupled* or *combined* with one another. *Coupling* of two operations implies that neither of them exists independently, only the application

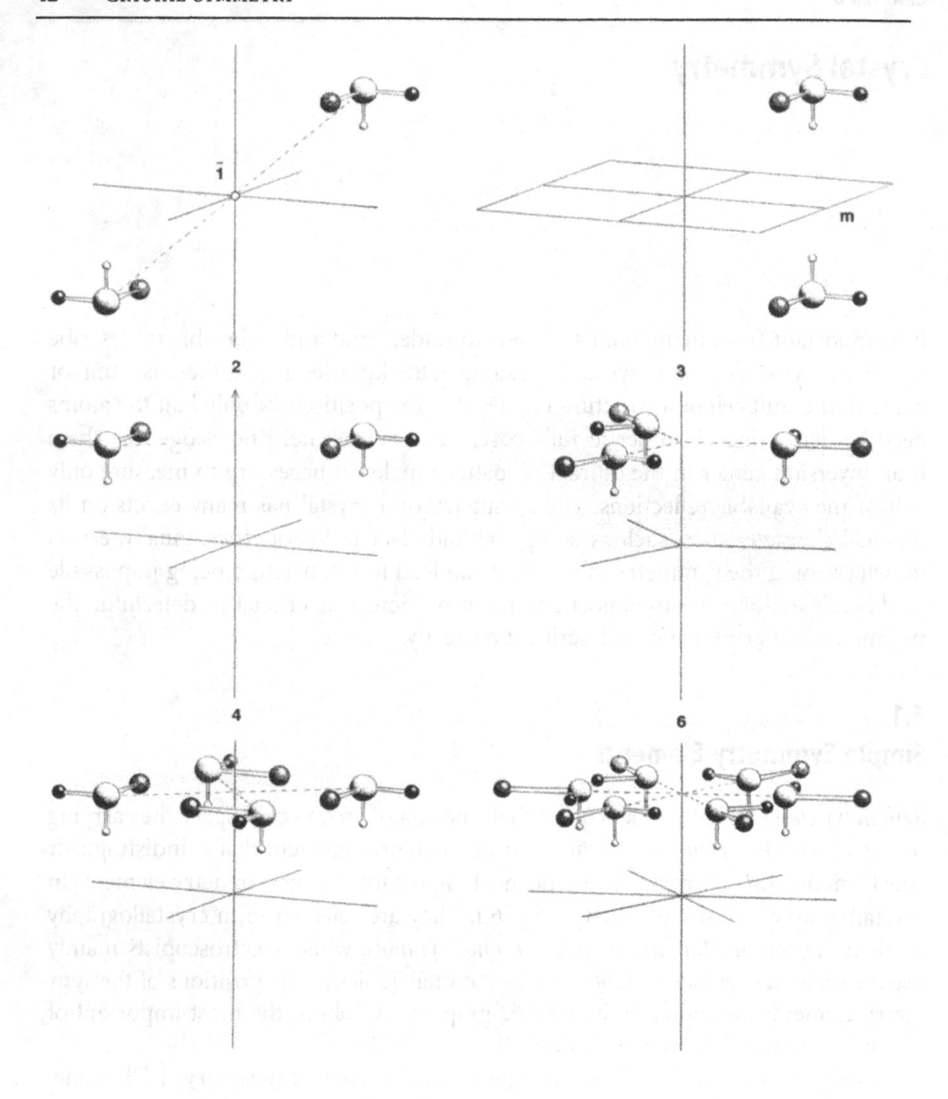

Fig. 6.1. Simple symmetry elements: $\bar{1}$ inversion center, *m* mirror plane, 2-, 3-, 4-, and 6-fold rotation axes.

of both of them together is a symmetry element. *Combination*, on the other hand, means that both elements and their combination exist independently.

6.1.1 Coupling of Symmetry Elements

The coupling of a rotation axis with an inversion center is called a rotoinversion axis. In the case of a 2-fold axis, this results in a mirror plane normal to the axis and consequently no new element (see also section 6.1.2). New rotoinversion axes do arise, however, on the coupling of 3-, 4-, and 6-fold axes with an inversion center to give the new $\bar{3}$-, $\bar{4}$- and $\bar{6}$-axes. For example, the operation of a $\bar{4}$-axis ("bar four" or "four bar" axis), which is characteristic of a tetrahedron distorted along a 2-fold axis, is shown in Fig. 6.2. A motif **1**, e.g. a ligand molecule, is rotated by the coupled 4-fold axis through 90° to **2**. The molecule in this position is *not* observed, but after inversion, its mirror image at **3** is. Another operation of the 4-fold axis places the molecule at **4**, which again is not observed, but a further inversion leads to a molecule at **5**. The same procedure leads via **6** to **7** and finally via **8** back to the original position **1**.

An analogous procedure included in Fig. 6.2 as well will lead to $\bar{3}$- and $\bar{6}$-axes. It is important to note that the 4-fold axis used to generate $\bar{4}$ is actually reduced to a 2-fold, and the 6-fold axis used to generate $\bar{6}$ is reduced to a 3-fold. In neither of these cases is an inversion center present. In the case of the $\bar{3}$-axis, coupling actually has the effect of a combination of symmetry elements (see the next section).

Using sketches like Fig. 6.2, it is easy to demonstrate that other couplings do not lead to new symmetry elements (in all of these examples, the coupled m is taken to be normal to the axis): $m + \bar{1} \rightarrow 2; 2 + m \rightarrow \bar{1}; 3 + m \rightarrow \bar{6}; 4 + m \rightarrow \bar{4}; 6 + m \rightarrow \bar{3}$ (note that rotoreflection is equivalent to rotoinversion, although not necessarily of the same type!)

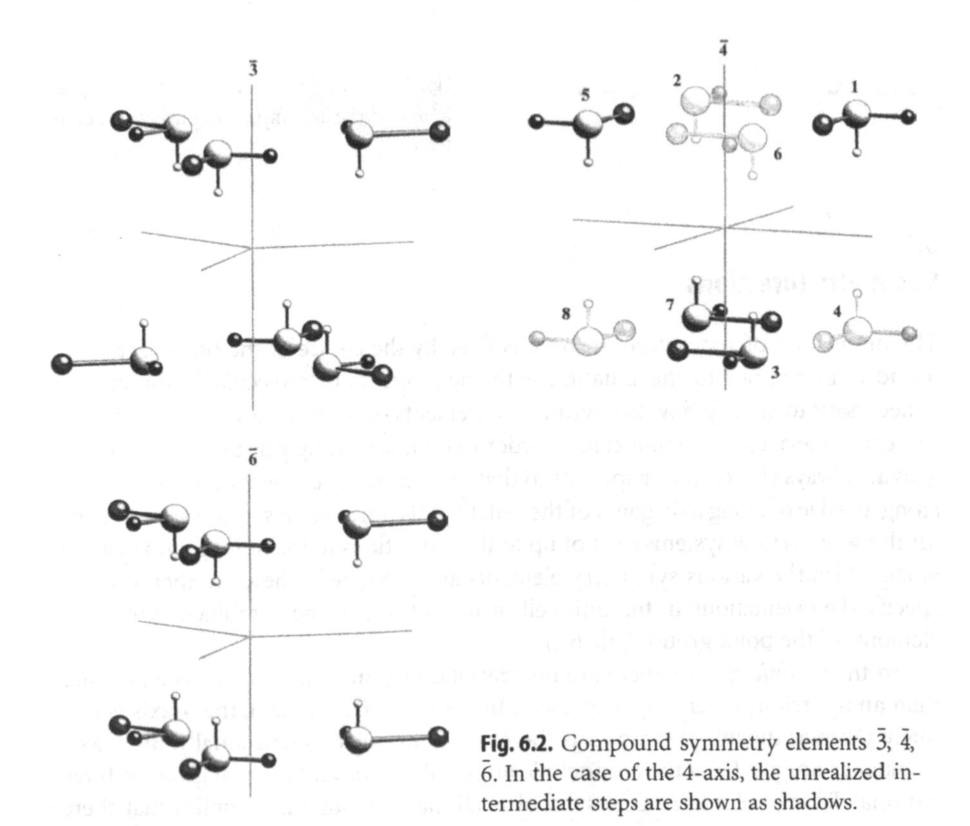

Fig. 6.2. Compound symmetry elements $\bar{3}$, $\bar{4}$, $\bar{6}$. In the case of the $\bar{4}$-axis, the unrealized intermediate steps are shown as shadows.

6.1.2
Combination of Symmetry Elements

In contrast to coupling, combination of symmetry elements implies the addition of two actual elements of the system. For example, the combination of a 2-fold axis with a mirror plane perpendicular to it is indicated by the symbol $2/m$ ("two over m").

Such a combination may give rise to other symmetry elements. In the case of $2/m$, (Fig. 6.3) an inversion center is generated at the point of intersection of the 2-fold axis and the mirror plane. Such generated elements are not usually indicated in the symbol, but their presence can be very important. This matter will be developed further in section 6.4. With the help of group theory, it can be shown that for crystals precisely 32 distinct combinations of symmetry elements can be made. These 32 crystallographic point groups or *crystal classes* will be discussed further in section 6.5.1.

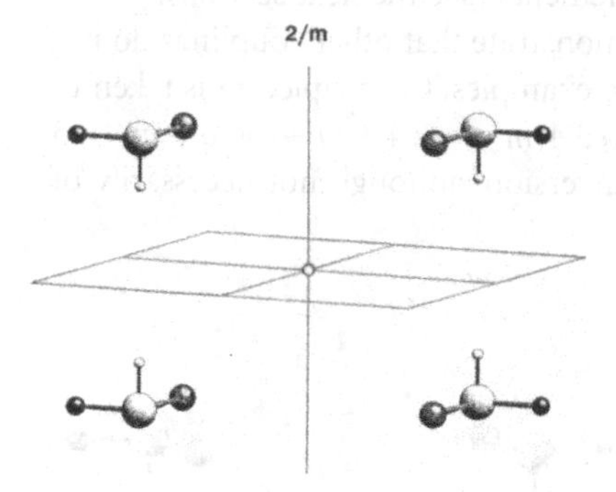

Fig. 6.3. Combination of 2 and m to give $2/m$ and the accompanying inversion center $\bar{1}$.

6.2
Symmetry Directions

The orientation of a structure in space is fixed by the choice of the basis vectors ***a***, ***b*** and ***c***. In contrast to the situation with the elements of molecular symmetry, it is necessary to specify how the symmetry elements of a crystal are orientated with respect to the axes of the unit cell. In order to do this as simply as possible, unit cell axes are always chosen (cf. chapter 2) so that the symmetry elements always lie either along an edge or along a diagonal of the cell. For this purpose, it is necessary to define for the seven crystal systems a set of up to three positions in the point group symbol along which the various symmetry elements are orientated. These are then used to specify the orientations in the unit cell of up to three of the combined symmetry elements of the point group (Tab. 6.1).

In the *triclinic* system, there are no special directions, since no symmetry other than an inversion center may be present. In the *monoclinic* system, the ***b***-axis is normally taken as the unique axis (β is the monoclinic angle). Occasionally, the ***c***-axis is taken as unique, (γ is the monoclinic angle). Both conventions are given in International Tables, Volume A. One symbol is all that is required: 2 implies that there

Table 6.1. Directions implied in the symmetry symbols for the seven crystal systems.

Crystal System	Order of directions	Examples
triclinic	—	$1, \bar{1}$
monoclinic	b	$2, 2/m$
orthorhombic	a, b, c	$mm2$
tetragonal	$c, a, [110]$	$4, \bar{4}, 4/mmm$
trigonal	$c, a, [210]$	$3, \bar{3}m1, 31m$
hexagonal	$c, a, [210]$	$6/m, \bar{6}2m$
cubic	$c, [111], [110]$	$23, m\bar{3}m$

is a 2-fold axis parallel to ***b***, *m* that there is a mirror plane *perpendicular* to ***b***, and $2/m$ that both elements are present. In the *orthorhombic* system, the symmetry along all three mutually perpendicular directions is given, in the order ***a***, ***b***, ***c***, e.g. $mm2$. For the *tetragonal* system, the first term in the symmetry symbol gives the 4-fold symmetry element along the unique ***c***-direction. Then, if there is one, follows an element along the ***a***-direction, which, because of the 4-fold symmetry, must apply to the ***b***-direction as well. In the third position comes any symmetry element along the ***a***, ***b***-diagonal, the $[110]$, and also the $[1\bar{1}0]$ direction. The *trigonal* and *hexagonal* systems are similar. First comes the element in the ***c***-direction, 3-fold in the trigonal system and 6-fold in the hexagonal, next, if necessary, the symmetry along $\boldsymbol{a}(=\boldsymbol{b})$ and finally along the $[210](=[\bar{1}10]=[\bar{1}\bar{2}0])$ diagonals, which are normal to the (100), (010) and $(\bar{1}10)$ planes, respectively, and consequently are the corresponding reciprocal axis-directions (Fig. 6.4). In the trigonal system, in addition to the 3-fold element along ***c***, there is often a symmetry element along *either* the real *or* the reciprocal axes in the other directions. To show this, a symbol is written either $3m1$ or $31m$, where the "1" means no symmetry. Unfortunately in the literature "$3m$" alone is sometimes written, making it unclear which of these distinct possibilities is meant.

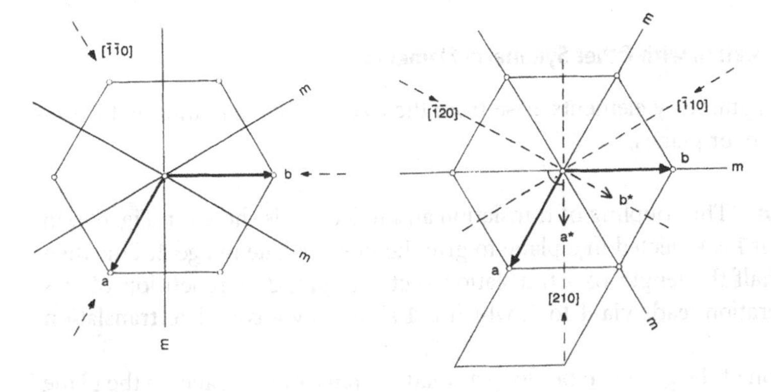

Fig. 6.4. Alternative symmetry directions in trigonal and hexagonal crystal classes shown for the case of symmetries $3m1$ (left) and $31m$ (right).

This abbreviation is, however, permissible for R-lattices, where symmetry along [210] is never possible.

In the *cubic* system, the first direction given is along the axis $\boldsymbol{c}(=\boldsymbol{a}=\boldsymbol{b})$ which is either a 2-fold or 4-fold axis and/or a plane perpendicular to them. The second direction gives the 3-fold symmetry along the body diagonals ([111], $[\bar{1}11]$, $[1\bar{1}1]$ and $[11\bar{1}]$). If required, a third symbol gives the symmetry along the face diagonals ([110], [101], [011], $[\bar{1}10]$, $[\bar{1}01]$ and $[0\bar{1}1]$). Given the full symbol, it is easy to deduce the implied axial system. For example, if a 3-fold element is in the first place, the crystal must be trigonal, while if it is in the second place, the crystal is cubic. The directions of the symmetry elements have far-reaching effects on the three dimensional periodicity of the crystal.

6.3 Symmetry Elements Involving Translation

Molecular symmetry (point symmetry) may be described in terms of a single point, usually the center of gravity of the molecule. In the case of crystals, the translation along the basis vectors must also be considered. Lattice translation is a simple symmetry element itself, as it meets the conditions of the definition given in section 6.1. In addition, it may be combined and coupled with other symmetry elements.

6.3.1 Combination of Translation with Other Symmetry Elements

The *combination* of translation symmetry with an inversion center at the vertex of a unit cell is shown in Fig. 6.5. One result is, of course, that at each other vertex an identical inversion center appears. In addition, inversion centers are generated at the center of each edge and each face of the cell as well as at the body center. Similar results occur with mirror planes, and rotation and rotoinversion axes.

6.3.2 Coupling of Translation with Other Symmetry Elements

Important new symmetry elements arise from the *coupling* of translation with rotation axes and mirror planes.

Glide reflection The coupling of translation and reflection is shown in Fig. 6.6. In this case, a motif **1** is reflected in a plane to give the intermediate image **2**. This then shifts (glides) half the length of a translation vector to give **3**. A repetition of this compound operation leads via **4** to **5**, which is **1** shifted by a complete translation vector.

The direction of the glide can be along any lattice translation parallel to the plane of reflection, which is then called a *glide plane a, b,* or *c*. The glide can also be along a face diagonal (*diagonal glide plane*) when it has the symbol n. Which plane is meant

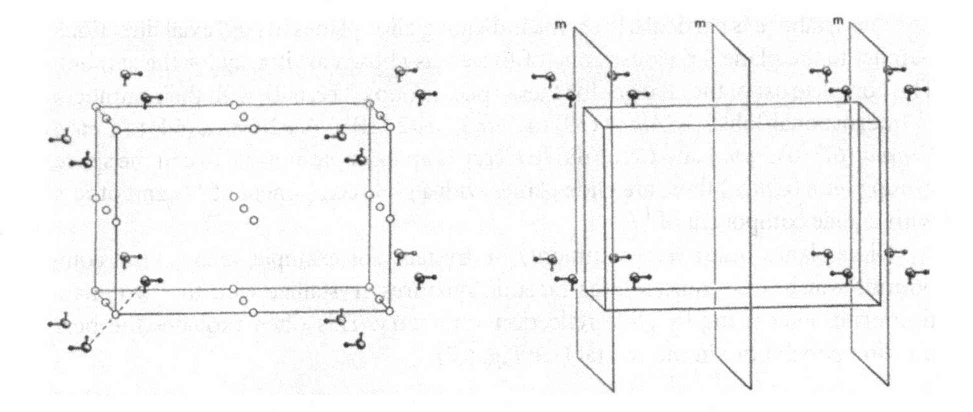

Fig. 6.5. Combination of translation with an inversion center (left) and a mirror plane (right).

is indicated by the position of the symmetry element in the symmetry symbol (cf. section 6.2). If, for example, the symbol refers to the ***b***-direction, the glide component must be $^{a}/_{2}+^{c}/_{2}$. If the face is centered (e.g. in a Bravais lattice of type F), this means that a translation vector from the origin ends there, the glide component will be only ¼ of the diagonal, and the operation sketched above must be repeated to traverse the entire diagonal. The symbol for this special diagonal glide plane is *d*, and as a typical example of it occurs in the cubic *F*-centered diamond structure, it is called a *diamond glide plane.* A summary of all possible glide planes is given in Table 6.4 (section 6.6.2) where it is also shown how they may be detected from the diffraction pattern of a crystal.

In a few centered space groups (see section 6.4 ff.) combinations of glide planes arise which cannot be described simply and completely by the symbols introduced thus far. These "double glide planes" have now been given the symbol *e* by a committee of the International Union of Crystallography (see International Tables A (1995)). A general symbol *g* has also been proposed for a few cases which are ambiguous in terms of the Hermann-Mauguin symbol. (See International Tables A, section 11.2 (1995)). Although these symbols are not yet in universal use, their application is highly recommended.

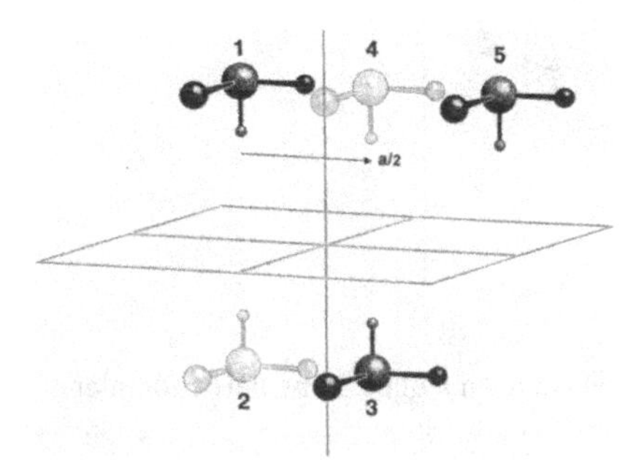

Fig. 6.6. Glide reflection: the unrealized intermediate steps are shown as shadows.

The symbol *e* is particularly useful, indicating glide planes in *both* axial directions parallel to the plane. Previously, one of these was chosen arbitrarily for the symbol. The complete list of the changes for these space groups (Sec. 6.4), with their numbers in International Tables, is: *Abm*2 (39) to *Aem*2, *Aba*2 (41) to *Aea*2, *Cmca* (64) to *Cmce*, *Cmma* (67) to *Cmme*, and *Ccca* (68) to *Ccce*. As an example, normal to *c* in the space group *Cmce* (*Cmca*) there are glide planes with a glide component of $^a/_2$ and others with a glide component of $^b/_2$.

Glide planes occur very commonly in crystals. For example, many chiral compounds which are synthesized as racemic mixtures crystallize with the two enantiomorphs alternating by glide reflection symmetry. This often provides the best packing possibilities in the crystal (see Fig. 6.7).

Screw axes The coupling of translation with a rotation axis leads analogously to the so-called *screw axes.* For any rotation axis of order *n*, there are $n - 1$ modes of coupling with translation in the direction of the axis: the motif is rotated by $360°/n$, and immediately translated m/n of the lattice translation along the axis, where *m* is an integer between 1 and $n - 1$. The value of *m* is then appended to the symbol for the rotation axis as subscript: n_m.

For the 2-fold axis, the only possibility is thus the 2_1-axis (two-one axis) (Fig. 6.8), an exceptionally common symmetry element. Like the glide plane, it gives good packing possibilities, and, as it merely rotates the motif by 180°, it does *not* invert the configuration of a molecule and so is suitable for packing molecules of a single enantiomer.

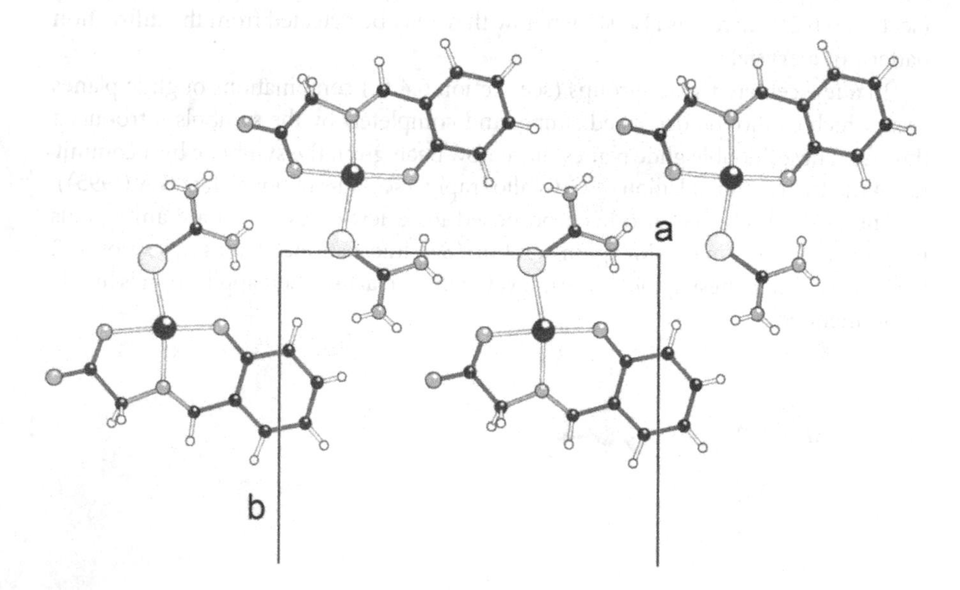

Fig. 6.7. Example of molecular packing via glide plane (here an a glide plane perpendicular to the *b*-axis).

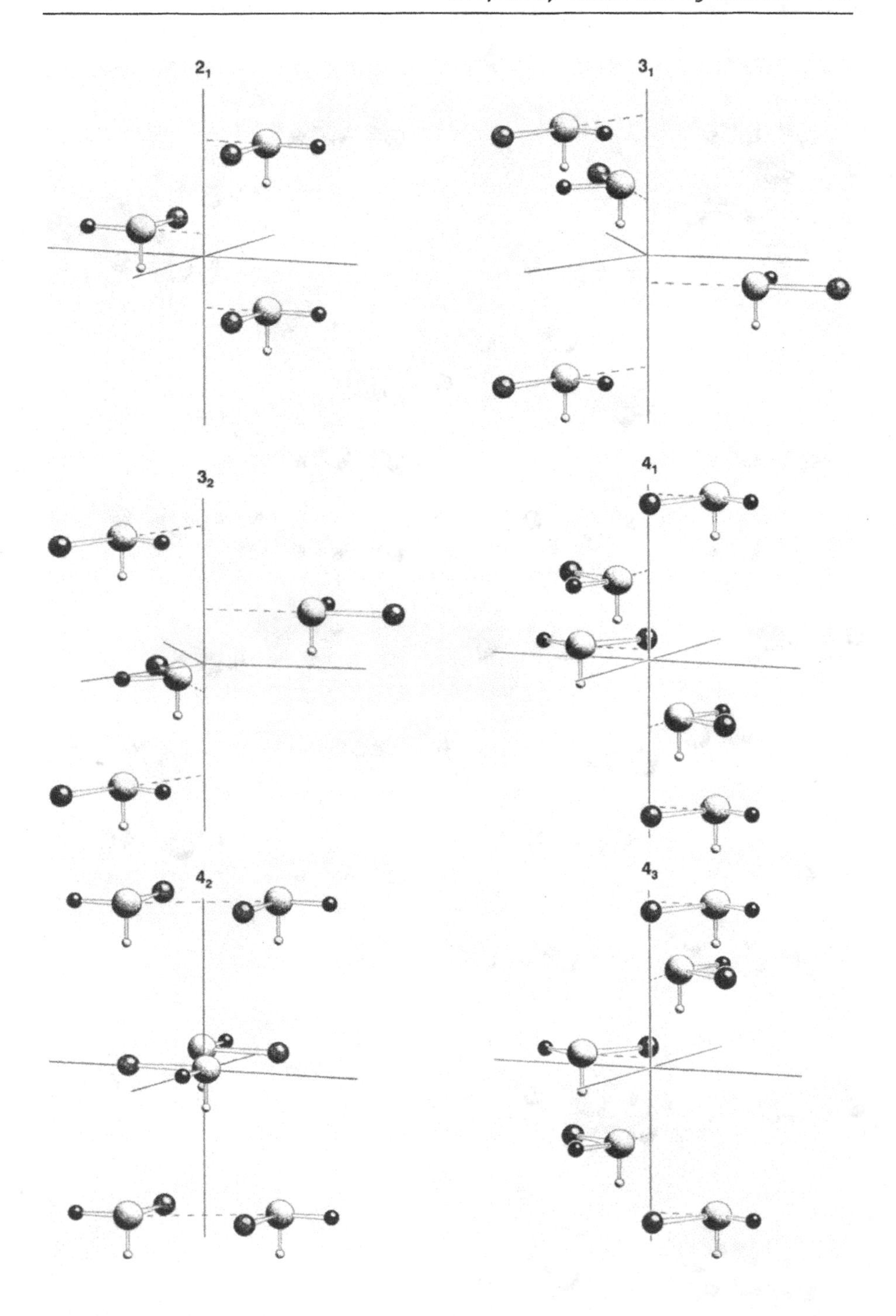

Fig. 6.8. Screw axes: two-fold screw axes 2_1; three fold screw axes 3_1, 3_2, four-fold screw axes 4_1, 4_2, 4_3; and six-fold screw axes 6_1, 6_2, 6_3, 6_4, 6_5.

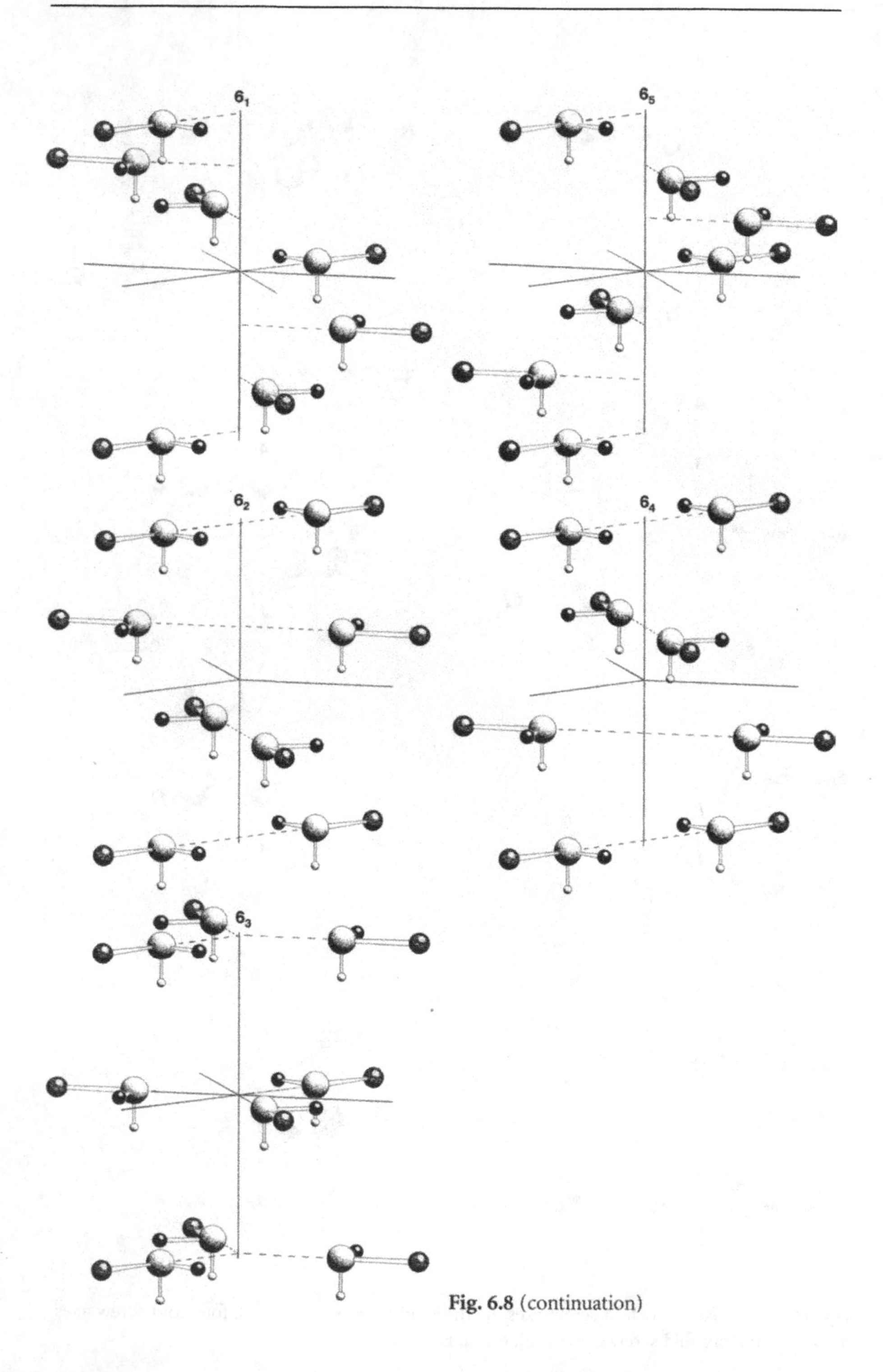

Fig. 6.8 (continuation)

The effect of the subscript m on the sense of screw — and why the term *screw* is used at all — is made clear by consideration of the 3-fold screw axes. A 3_1-axis implies a *clockwise* rotation of $360°/n = 120°$, followed by a translation of $m/n = ⅓$ of the lattice constant in the direction of the axis, usually c. The motif is thus repeated at $0°$, $z = 0$; $120°$, $z = ⅓$; $240°$, $z = ⅔$; $360°$, $z = 1$, etc., all in the direction of a *right handed screw*. For a 3_2-axis, on the other hand, the translation component will be ⅔ of the c-axis. Since a whole number of translations may always be added to or subtracted from any position, the motif will now occur at $0°$, $z = 0$; $120°$, $z = ⅔$; $240°$, $z = 4/3$ and ⅓; $360°$, $z = 6/3 = 2$ and 1, etc. This time, a *left handed screw* has been described. By the same procedure, the several variants of 4- and 6-fold axes may be derived (Fig. 6.8). These are seldom encountered in real crystals.

That then concludes the list of all the symmetry elements that are significant in crystallography. In Table 6.2, are displayed the graphical symbols used for these symmetry elements in International Tables A. They are useful when it is desired to include the position of symmetry elements in a graphical representation of the unit cell.

Table 6.2. Important graphical symbols for symmetry elements. * axis in the plane of the paper, # glide direction in plane of paper, & glide direction normal to plane of paper.

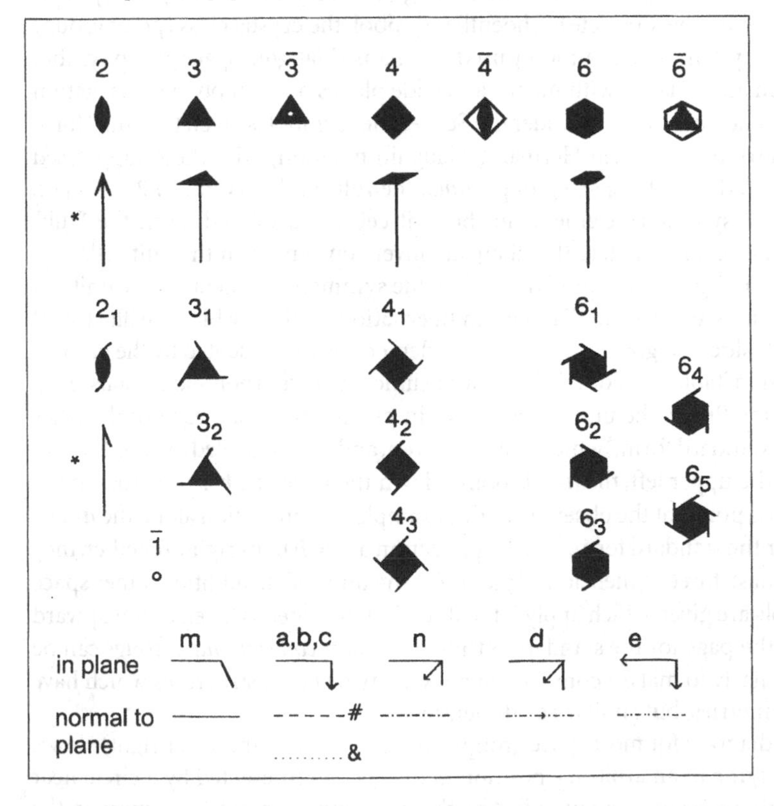

6.4
The 230 Space Groups

It is possible, using group theory, to determine all the possible combinations of the simple and compound symmetry operations, including those involving translation. There are precisely 230 of these, the 230 *space groups*. They determine the types and positions of symmetry elements that are possible for a crystal structure.

6.4.1
Space-group Notation in International Tables for Crystallography

Every crystal may be described in terms of its space group, which are compiled in International Tables A, each with a serial number, and arranged more or less in order of increasing complexity. Since all crystallographers accept the conventions of these tables, the comparison and exchange of structure data by different workers is made much simpler. In what follows, the space group *Pnma* (№ 62) is used as an example (Fig. 6.9).

Space-group symbol: The space-group symbol (upper left corner) always begins with a capital letter, giving the Bravais lattice type. This is followed by the symbols for the chief symmetry operations in the order given in section 6.2 for point groups. Next are given the (now obsolete) Schoenflies symbol, the crystal class (point group) and the crystal system. The commonly used Hermann-Mauguin space group symbol is, in fact, often abbreviated, with mirror and glide planes often implying the rotation and screw axes normal to them. Under the Schoenflies symbol is given the "full" form of the space group symbol (in Hermann-Mauguin notation) with these suppressed elements included. For the space group *Pnma*, the full symbol is $P2_1/n\ 2_1/m\ 2_1/a$. Even so, not all symmetry elements in the unit cell are displayed. Even the "full" symbol does not explicitly state that there are inversion centers in the unit cell.

Space-group diagrams: A summary of all of the symmetry elements in the unit cell and their positions are given in diagrams in International Tables, which are illustrated here on the left side of Fig. 6.9. The symmetry elements are represented by the graphic symbols given in Table 6.2. For triclinic, monoclinic and orthorhombic crystals, as in this example for *Pnma*, the unit cell is shown in projections along each of the basis vectors. The "standard form," which is always given, and has unlabeled axes, is that the origin lies in the upper left, the ***a***-axis points down the page, the ***b***-axis across to the right, the ***c***-axis points at the observer. In this example, the projection along the ***a***-axis is given under the standard form, and the projection along ***b*** to its right. Together, they make up a "quasi-three-dimensional" picture of the unit cell. In addition, other space group symbols are given which apply if the direction indicated is taken as the upward direction on the page for the standard setting. Such *non-conventional settings* can be useful, particularly to make a comparison between two crystal structures which have different symmetries, but similar cell dimensions.

The final diagram for most space groups (in Fig. 6.9 that in the lower right) shows how an atom lying in an arbitrary position in the cell — represented by a circle near the origin with a +-sign — is multiplied by the symmetry elements. A comma in the

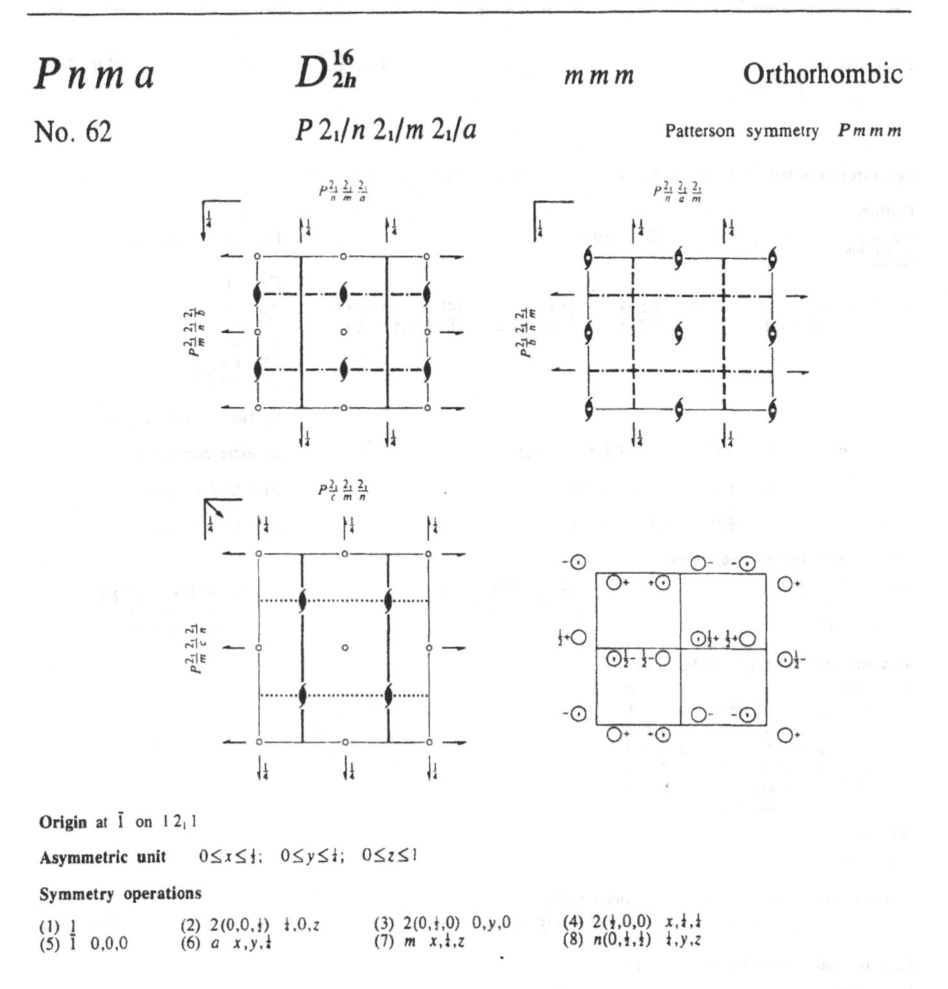

Pnma
D_{2h}^{16}
mmm
Orthorhombic
No. 62
$P\,2_1/n\,2_1/m\,2_1/a$
Patterson symmetry Pmmm
Origin at $\bar{1}$ on $1\,2_1\,1$
Asymmetric unit $0 \le x \le \frac{1}{2};\ 0 \le y \le \frac{1}{4};\ 0 \le z \le 1$
Symmetry operations
(1) 1
(2) $2(0,0,\frac{1}{2})\ \frac{1}{4},0,z$
(3) $2(0,\frac{1}{2},0)\ 0,y,0$
(4) $2(\frac{1}{2},0,0)\ x,\frac{1}{4},\frac{1}{4}$
(5) $\bar{1}$ 0,0,0
(6) $a\ x,y,\frac{1}{4}$
(7) $m\ x,\frac{1}{4},z$
(8) $n(0,\frac{1}{2},\frac{1}{2})\ \frac{1}{4},y,z$

Fig. 6.9. Diagram for space group *Pnma* extracted from *International Tables for Crystallography, Vol. A* (with permission of Kluwer Academic Publishers). Continuation on the next page.

circle implies that it is the mirror image of the original atom + and − indicate that the atom lies above or below the plane of the paper. In addition, any translation in the z-direction is shown. For example, $\frac{1}{2}$ implies a translation of $\mathbf{c}/2$. It can be seen that in space group *Pnma*, each atom is expanded by the symmetry operations into eight. If the position of any one is known, the other seven follow automatically from the symmetry.

Positions: Allowing each symmetry operation in turn to operate algebraically on a point in what is called a *general position,* with coordinates xyz, the result is the *general equivalent positions,* here eight in number. Thus, this is said to be a space group with a multiplicity of 8, as it has 8-fold general positions. The diagrams in Fig. 6.9 may be used to make these points clear. For example, position (2): $-x + ½, -y, z + ½$, is the operation of the 2_1-axis parallel to $\mathbf{c}$ at 0, ¼, z. Similarly, position (5): $-x, -y, -z$ is

CONTINUED **No. 62** ***Pnma***

Generators selected (1); $t(1,0,0)$; $t(0,1,0)$; $t(0,0,1)$; (2); (3); (5)

Positions

Multiplicity, Wyckoff letter, Site symmetry	Coordinates				Reflection conditions
					General:
8 d 1	(1) x,y,z	(2) $\bar{x}+\frac{1}{2},\bar{y},z+\frac{1}{2}$	(3) $\bar{x},y+\frac{1}{2},\bar{z}$	(4) $x+\frac{1}{2},\bar{y}+\frac{1}{2},\bar{z}+\frac{1}{2}$	$0kl$: $k+l=2n$
	(5) $\bar{x},\bar{y},\bar{z}$	(6) $x+\frac{1}{2},y,\bar{z}+\frac{1}{2}$	(7) $x,\bar{y}+\frac{1}{2},z$	(8) $\bar{x}+\frac{1}{2},y+\frac{1}{2},z+\frac{1}{2}$	$hk0$: $h=2n$
					$h00$: $h=2n$
					$0k0$: $k=2n$
					$00l$: $l=2n$
					Special: as above, plus
4 c $.m.$	$x,\frac{1}{4},z$	$\bar{x}+\frac{1}{2},\frac{3}{4},z+\frac{1}{2}$	$\bar{x},\frac{3}{4},\bar{z}$	$x+\frac{1}{2},\frac{1}{4},\bar{z}+\frac{1}{2}$	no extra conditions
4 b $\bar{1}$	$0,0,\frac{1}{2}$	$\frac{1}{2},0,0$	$0,\frac{1}{2},\frac{1}{2}$	$\frac{1}{2},\frac{1}{2},0$	hkl : $h+l,k=2n$
4 a $\bar{1}$	$0,0,0$	$\frac{1}{2},0,\frac{1}{2}$	$0,\frac{1}{2},0$	$\frac{1}{2},\frac{1}{2},\frac{1}{2}$	hkl : $h+l,k=2n$

Symmetry of special projections

Along [001] $p2gm$
$a'=\frac{1}{2}a$ $b'=b$
Origin at $0,0,z$

Along [100] $c2mm$
$a'=b$ $b'=c$
Origin at $x,\frac{1}{4},\frac{1}{4}$

Along [010] $p2gg$
$a'=c$ $b'=a$
Origin at $0,y,0$

Maximal non-isomorphic subgroups

I	$[2]P2_12_12_1$	1; 2; 3; 4
	$[2]P112_1/a\,(P2_1/c)$	1; 2; 5; 6
	$[2]P12_1/m1\,(P2_1/m)$	1; 3; 5; 7
	$[2]P2_1/n11\,(P2_1/c)$	1; 4; 5; 8
	$[2]Pnm2_1\,(Pmn2_1)$	1; 2; 7; 8
	$[2]Pn2_1a\,(Pna2_1)$	1; 3; 6; 8
	$[2]P2_1ma\,(Pmc2_1)$	1; 4; 6; 7
IIa	none	
IIb	none	

Maximal isomorphic subgroups of lowest index

IIc $[3]Pnma\,(a'=3a)$; $[3]Pnma\,(b'=3b)$; $[3]Pnma\,(c'=3c)$

Minimal non-isomorphic supergroups

I none

II $[2]Amma\,(Cmcm)$; $[2]Bbmm\,(Cmcm)$; $[2]Ccmb\,(Cmca)$; $[2]Imma$; $[2]Pnmm\,(2a'=a)\,(Pmmn)$; $[2]Pcma\,(2b'=b)\,(Pbam)$; $[2]Pbma\,(2c'=c)\,(Pbcm)$

the inversion center at the origin. The eight coordinate triples thus give the symmetry operations in terms of their locations in the unit cell.

A special situation arises when an atom falls precisely on a symmetry element without a translation component; in this example, a mirror plane or an inversion center. On such *special positions,* at least one atomic parameter must be fixed, for example y must be exactly ¼ or ¾ to lie on a mirror plane in *Pnma*; for the inversion centers, 0, 0, 0; 0, 0, $\frac{1}{2}$, etc. all parameters are fixed. Since an atom, lying on a symmetry element, is left unchanged by that element, the usual doubling does not occur, and the multiplicity of this position is less than eight. In *Pnma,* for example, the special positions are all 4-fold. This multiplicity is given in the first column of the table of coordinates (Fig. 6.9, right hand side). All the possible position types for the space group, beginning with the "most special" and ending with the general position are given an alphabetical symbol, which is called the *Wyckoff symbol* for the position. Normally, this symbol is used along with the multiplicity, and a position is described as, e. g. "the 4*c* position" in *Pnma*. The symmetry element corresponding to the special position is given in column 3 in a way that indicates its orientation, as in section 6.2. For example, the symbol ".*m*." for position 4*c* indicates that it lies on a mirror plane normal to ***b***. The knowledge that a special position is occupied can give important information about a structure. If, for example, the central atom of a complex lies on position 4*c* in space group *Pnma*, that implies that the complex must have the corresponding point symmetry, here m (C_s).

If the space group has a centered Bravais lattice, the general translation operations are given before the general position, e. g. for *C*-centering $(\frac{1}{2}, \frac{1}{2}, 0)+$. This means that for the general position as well as every special position, another atom with the coordinates $(x + \frac{1}{2}, y + \frac{1}{2}, z)$ is generated as well. Further information in Fig. 6.9, will be referred to later.

Choice of Origin. The origin of the unit cell is chosen so that it occupies a position of relatively low multiplicity. In some space groups, mostly of higher symmetry, there are two possibilities for doing this, and they naturally lead to different locations for the special positions. In such cases, International Tables often gives both choices. In space group *Fddd*, for example the origin may be located either in the special position 8*a*, with symmetry 222, or in 16*c*, an inversion center. In such cases, the origin should always be chosen to lie on the inversion center, as this gives great advantages for both the solution and the refinement of the structure. (cf. section 8.3) In practice, then, it is important to be certain that the "correct," centrosymmetric choice has been made.

6.4.2 Centrosymmetric Crystal Structures

The advantage of centrosymmetric structures is that the expression for the structure factor is greatly simplified. When the origin lies on an inversion center, for every atom with parameters xyz, there is a corresponding one at $\bar{x}\bar{y}\bar{z}$. Summing the contributions of both these atoms to the structure factor equation then gives:

$$\begin{aligned} F_c(hkl) &= f\cos[2\pi(hx + ky + lz)] + fi\sin[2\pi(hx + ky + lz)] \\ &\quad + f\cos[-2\pi(hx + ky + lz)] + fi\sin[-2\pi(hx + ky + lz)] \end{aligned}$$

changing the sign of the argument of the cosine does not change its value so they reinforce, but the sine terms cancel, leaving only the real part:

$$F_c(hkl) = 2f\cos[2\pi(hx + ky + lz)]$$

The structure factor is thus no longer a complex quantity, and the phase problem is reduced to the problem of whether the sign of F is positive or negative.

6.4.3 The Asymmetric Unit

The *asymmetric unit* is the name given to that minimum group of atoms whose positions, together with those generated by the symmetry operations of the space group generate the complete contents of the unit cell. In structures composed of molecules of a single type which have no symmetry themselves, the asymmetric unit is usually a molecule, a "formula unit," or occasionally two or more molecules which differ from one another in orientation or conformation. When a molecule has symmetry which can conform to crystallographic symmetry, it may occupy a special position, and the asymmetric unit will then be a half molecule or even some smaller fraction. Sometimes, two independent half molecules can constitute the asymmetric unit, e. g. when two molecules occupy two inversion centers in the space group $P2_1/c$. In inorganic solid-state materials, e. g. coordination compounds forming 3-dimensional networks, such as the perovskites ABX_3, the asymmetric unit may be such a formula unit, a fraction of it or an integral multiple of it. Since the asymmetric unit is reproduced by the symmetry elements, the unit cell must invariably contain an integral multiple of the asymmetric unit.

Number of formula units per unit cell, Z: It is useful at the beginning of a structure determination, to know Z, the number of formula units in the cell, since when the space group is known, it is then possible to calculate how big the asymmetric unit is, i. e. how many atoms must have coordinates found for them! Knowing Z, it is possible to express the density of the substance as in equation 6.1:

$$d_c = \frac{M_r \cdot Z}{V_{EZ} \cdot N_A} \tag{6.1}$$

The "X-ray density" d_c of a material may be thus calculated from the mass of a formula unit (M_r/N_A) where M_r is the molar mass and N_A is the Avogadro number, the number of formula units per cell, Z, and the volume of the unit cell, V, calculated from the lattice constants. If the measured density, which is usually slightly less than this ideal value, is substituted for it in the above equation, an unknown Z may be calculated. Since Z must be an integer, and usually a small one, it often suffices to use the density of some analogous compound in order to obtain the right value of Z.

Still simpler is the use of a rule of thumb relating to the average volume occupied by an atom in a solid [19]. Assigning an approximate volume of 17 Å^3 to each of the M non-hydrogen atoms in the formula, Z is simply

$$Z = \frac{V_{EZ}}{17 \cdot M} \tag{6.2}$$

Comparison of Z with the multiplicity of the general position n_a of the space group will then require that the asymmetric unit is Z/n_a times the formula unit. The goal of a structure determination is to determine the positions of all atoms in the asymmetric unit.

In International Tables, an asymmetric unit is given for each space group. That for *Pnma* is given in Fig. 6.9. It gives the fraction of the unit cell volume which, by application of the symmetry operations generates the entire cell. It is possible to fit the asymmetric unit precisely into this volume. Usually this is not done, as it is more useful to describe an asymmetric unit as a molecule or half molecule of sensibly connected atoms, whether these atoms happen to occupy that specific volume or not. This is, of course, permissible, since *any* symmetry equivalent position for an atom is equivalent to any other. In the group-theoretical sense, space groups are closed groups. This implies that whichever equivalent position is chosen for any of the atoms, the application of the symmetry operations will define the same overall array of atoms.

6.4.4 Space Group Types

In the description of the symmetry of a crystal, the unit cell and its measurement are fundamental. The space group indicates which symmetry elements are present in the unit cell and where they are. Each crystal structure thus has a characteristic space group as it has a characteristic unit cell. Two different structures which are both described in *Pnma*, are said to crystallize in the same space group type, not in the same space group. There are even cases in which one and the same substance has two distinct structural modifications which nonetheless have the same space group type. (e.g. $MnF_3 \cdot 3H_2O$ [20]).

6.4.5 Group-Subgroup Relationships

For better understanding structural correlations or structural alterations connected with phase changes, the group-theoretical relationships between space groups can be very useful. They have been particularly explored by Bärnighausen [21], and a simple introduction with examples is also given by Müller [22]. For a particular family of related structures, that space group of highest symmetry within the family is identified as the *ideal type* or *aristotype*. From this, *maximal subgroups* are identified by removing single symmetry elements (and all of those that are present as a combination involving the removed symmetry element). Conversely, the aristotype is said to be a *supergroup* of its subgroups. The conversions which this deconstruction brings about result in a sort of "tree" (Fig. 6.10) and are of three types:

1. t-subgroups (from the German *translationsgleiche*) In this case, the unit cell is unchanged, but the removal of symmetry elements moves the structure into a lower crystal class. They are given an order of 2 or 3 depending on the factor by which the number of symmetry elements has been reduced, and the transformation is described as a "t-transformation of index 2 (or 3)" and written $t2$ or $t3$.
2. k-subgroups (from the German *klassengleiche*) The crystal class and the crystal system are unchanged, but either the Bravais lattice changes, as centering is lost, or the unit cell is doubled or trebled in size. In this case, the symbol is written as above $k2$ or $k3$, and the transformation of the lattice constants is given.

t-type	k-type	isomorphous
$P\frac{4}{n}\frac{2}{m}\frac{2}{m}$(129)	$C\frac{2}{m}$(12)	$P\frac{2_1}{a}$(14)
$CsFeF_4$		CuF_2
$t2$	$k2$	$i2$
		$c' = 2c$
↓	↓	↓
$P\frac{4}{n}$(85)	$P\frac{2_1}{a}$(14)	$P\frac{2_1}{a}$(14)
$CsMnF_4$	$RbAuBr_4$	VO_2

Fig. 6.10. Examples for group-subgroup relationships (partially based on [21]).

3. Isomorphous subgroups. These are special types of k-groups, in which the space group type is unaltered. The enlargement of the lattice constants, however, results in the symmetry elements being "diluted".

It is important to note that it is certainly *not* possible to conclude that two structures have a structural relationship simply because they have space groups with a group-subgroup relationship. It is essential to be able to explain a mechanism whereby the atom positions of the less symmetrical structure can be derived from those of the more symmetrical. The necessary maximal subgroups for studying such transformations are given in International Tables (see Fig. 6.9, lower right). Important applications of them include twinning possibilities (cf. Section 11.2) and the discussion and understanding of structural transformations. Often, the correct space group for a variant structure of lower symmetry can be found by examining the choice of subgroups given for the ideal type. The relationship between groups is important, above all in the case of crystallographic phase changes: Only when a group-subgroup relationship exists can a second-order phase change occur.

6.5 Visible Effects of Symmetry

So far, the discussion of space groups has been concerned with the description of the symmetry properties of a crystal. At the beginning of a structure determination, these symmetry properties are unknown, and it must now be shown how these properties may be experimentally established.

6.5.1 Microscopic Structure

Space groups describe the geometric rules for the microscopic arrangement of atoms in a crystal. They include symmetry elements which include translation, and can only be made visible when a model or at least drawings of the structure are made. These symmetry elements, however, give rise to physical effects which can be observed experimentally, and these will be explained in the following sections.

6.5.2 Macroscopic Properties and Crystal Classes

Translation symmetry is irrelevant to most physical properties of crystals. It is, for example, important to know whether reflection symmetry is present, i.e. whether a chiral molecule is present with or without its enantiomorph. Whether the reflection does or does not have a glide component is irrelevant. Further, it is important whether a molecule is present in the four orientations associated with 4-fold rotation, but not whether this operation is a pure rotation or has a screw component. These observations are important for the optical and electrical properties of the crystal and for the external appearance, the *habit* of the crystal. The significant symmetry classes which describe these macroscopic properties are not the 230 space groups, with their translation-dependent symmetry. By the omission of the symbol for the Bravais lattice type, conversion of all glide planes into mirror planes *m*, and removal of all screw components from screw axes, the 230 space groups are reduced to the 32 point groups or *crystal classes* (Table 6.3). From the opposite point of view, each crystal class represents a number of space groups.

6.5.3 Symmetry of the Lattice

If the entire contents of the unit cell is ignored, leaving only the "bare" lattice, without the symmetry properties to regenerate the structure, only the symmetries of the 7 crystal systems remain. They define the metric symmetry of the unit cell.

If only the shapes of the uncentered unit cells are considered, the trigonal and hexagonal systems are indistinguishable, and many authors thus define only 6 crystal systems. However, the centered rhombohedral lattice is only trigonal (cf. Fig. 2.7, section 2.2). For this reason, International Tables separate all trigonal space groups into a separate trigonal system.

If the possible translation vectors that lead to centerings are also considered, the group of distinguishable lattices expands to the 14 Bravais lattices, already described in chapter 2.

6.5.4 Symmetry of the Diffraction Pattern – The Laue Groups

It was mentioned in chapter 4 that it is advantageous to represent the diffraction pattern of a crystal as a reciprocal lattice. If the points *hkl* of the reciprocal lattice are made to represent the measured diffraction intensities, this "weighted" reciprocal lattice effectively becomes a picture of the diffraction pattern of the crystal. It will then show how the symmetry of the crystal is related to the symmetry of the diffraction pattern, and which groups of reflections are "symmetry equivalent". The reflections are, as has been shown (section 3.4), directly related to the lattice planes, and since these are macroscopic features of the crystal, the expected symmetry, like that of the crystal faces, will only relate to the crystal class, not to the space group. There is a further difficulty: even if the crystal itself has no inversion center, its diffraction pattern, and consequently the weighted reciprocal lattice will always be centrosymmetric. This may be seen by comparing the structure factor expressions (equation 5.12, section 5.3) for two reflections *hkl* and $\bar{h}\bar{k}\bar{l}$ related to one another by inversion through the origin:

$$F(hkl) = \sum \{f_i \cos[2\pi(hx_i + ky_i + lz_i)] + i \sin[\ldots]\}$$

$$F(\bar{h}\bar{k}\bar{l}) = \sum \{f_i \cos[-2\pi(hx_i + ky_i + lz_i)] + i \sin[-\ldots]\}$$

Since $\cos\varphi = \cos -\varphi$ and $\sin -\varphi = -\sin\varphi$, these reflections will have the same amplitude, although the phase angle will have the opposite sign (Fig. 6.11). Since all that can be measured in the intensity is the square of the amplitude, the two intensities are equal, a relationship known as *Friedel's Law.* In fact there are important small deviations from Friedel's Law, which will be described further in section 10.2. Friedel's Law limits the symmetry of the diffraction pattern further: it must have the symmetry of a centrosymmetric crystal class. This means, for example, in the monoclinic crystal system, that a crystal in class 2 will have this combined with the $\bar{1}$ of the diffraction pattern so that it appears to be in class $2/m$. The same result will occur for crystals in class m. Thus it is not possible to distinguish the classes 2, m, and $2/m$ from their diffraction patterns. The same effect occurs in other crystal systems, and of the 32 crystal classes, only 11 distinct *Laue groups* remain. These are summarized in Table 6.3. In the tetragonal, trigonal, hexagonal and cubic systems, there are two Laue groups — one of higher symmetry than the other. From the diffraction pattern, it is possible to assign every crystal to one of these 11 Laue groups.

One special effect occurs for the more symmetrical trigonal Laue group $\bar{3}m$, where the mirror plane may have two distinct orientations relative to the crystal axes: It can be normal to the a-axis ($\bar{3}m1$), or normal to the [210] direction ($\bar{3}1m$) (cf. section 6.2). Thus, this Laue group is divided into two clearly distinguishable isomorphous groups, and the same argument holds for the subgroups 32 and $3m$. In practice then, it can

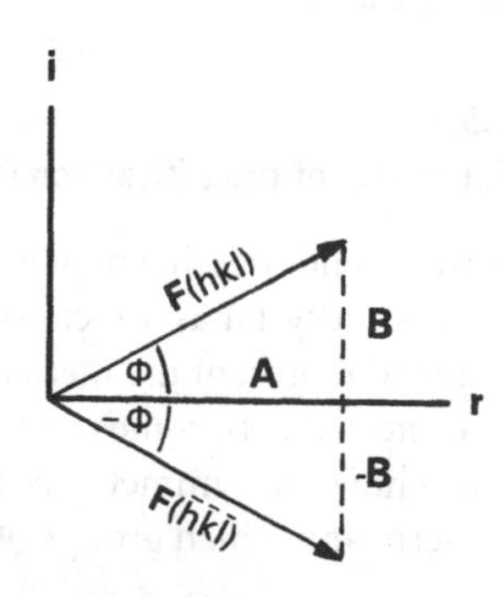

Fig. 6.11. Representation of structure factors for a Friedel-pair of reflections in the complex plane.

Table 6.3. The 32 crystal classes, distributed over the seven crystal systems and their Laue groups. (Isomorphous groups, with symbols referred to the crystallographic axes are given in brackets.)

Crystal System	Crystal class (Schönflies symbols in brackets)	Laue group
triclinic	$1\ (C_1), \bar{1}\ (C_i)$	$\bar{1}$
monoclinic	$2\ (C_2), m\ (C_s), 2/m\ (C_{2h})$	$2/m$
orthorhombic	$222\ (D_2), mm2\ (C_{2v}), mmm\ (D_{2h})$	mmm
tetragonal	$4\ (C_4), \bar{4}\ (S_4), 4/m\ (C_{4h})$	$4/m$
	$422\ (D_4), 4mm\ (C_{4v}), \bar{4}2m\ (D_{2d}), 4/mmm\ (D_{4h})$	$4/mmm$
trigonal	$3\ (C_3), \bar{3}\ (C_{3i})$	$\bar{3}$
	$\begin{bmatrix} 321 \\ 312 \end{bmatrix} (D_3), \begin{bmatrix} 3m1 \\ 31m \end{bmatrix} (C_{3v}), \begin{bmatrix} \bar{3}m1 \\ \bar{3}1m \end{bmatrix} (D_{3d})$	$\begin{bmatrix} \bar{3}m1 \\ \bar{3}1m \end{bmatrix}$
hexagonal	$6\ (C_6), \bar{6}\ (C_{3h}), 6/m\ (C_{6h})$	$6/m$
	$622\ (D_6), 6mm\ (C_{6v}), \bar{6}2m\ (D_{3h}), 6/mmm\ (D_{6h})$	$6/mmm$
cubic	$23\ (T), m\bar{3}\ (T_h)$	$m\bar{3}$
	$432\ (O), \bar{4}3m\ (T_d), m\bar{3}m\ (O_h)$	$m\bar{3}m$

be said that it is possible to assign 12 Laue symmetries. Although there is little effect on the diffraction symmetry, some other crystal classes (e.g. $\bar{4}2m$ and $\bar{6}2m$) have different possible orientations with respect to the crystallographic axes.

Those reflections which make up an "asymmetric unit" of the reciprocal lattice and thus carry all of the structural information are called the *independent reflections.* All other reflections may be generated from them by the symmetry operations of the Laue group, and they are called *symmetry equivalent reflections.* In fact, for a more accurate structure determination, it is worthwhile to measure some symmetry equivalent reflections, and then average them to give an averaged set of *independent reflections,* on the basis of which further calculation on the structure are made.

6.6 Determination of the Space Group

6.6.1 Determination of the Laue Group

The first step in space-group determination is the determination of the Laue symmetry. This is particularly straightforward if a set of single crystal photographs has been made, from which the intensity-weighted reciprocal lattice may be constructed (cf. Sect. 7.2). Program systems for four-circle diffractometers and area detector systems provide the possibility of representing color-coded measured intensities (cf. Sect 7.2 and 7.3) on a computer screen in the form of layers of the weighted reciprocal lattice.

For this to be effective, many pairs of possibly equivalent reflection must have been measured, so that it may be seen whether their intensities are equivalent within error limits or not. It is easy to choose suitable possibly equivalent reflections by drawing the symmetry elements of the Laue group onto a projection of the reciprocal lattice. (N.B. The directions corresponding to symmetry elements in the symbol refer to *direct* axes!). Thus, it is possible to see at a glance which reflections are related by these possible symmetry elements, and to test whether they do, in fact, have equivalent intensities. There are, of course, programs which do this job!

The handling of symmetry equivalent reflections is made easier in the trigonal and hexagonal systems by working with four Miller indices, $hkil$. The auxiliary index i gives the reciprocal axis intercepts on the $[\bar{1}\bar{1}0]$ axis, which is symmetry equivalent to a and b. It is calculated as $i = -(h+k)$. In this way reflections which are symmetry equivalent through rotation about the 3-fold axis normal to c have the simple cyclical interchange of hki indices: $hkil \longrightarrow kihl \longrightarrow ihkl$.

The determination to which of the 11 Laue groups a crystal belongs, and with it the determination of the crystal system, is only a small part of what is needed: for each Laue group there are a number of crystal classes, and for most crystal classes several space groups. It is necessary to obtain more symmetry information to narrow down the choice of possible space groups.

6.6.2 Systematic Absences

For this purpose, it is very useful that symmetry elements with translation components lead to the systematic cancelling or *absences* of certain reflections. These are sometimes called "extinctions," which is an unfortunate choice of word, as it also has a quite a different meaning, (cf. Sec. 10.3). How absences arise is clearly shown by the example of the pure translation symmetry of body-centering (Bravais lattice type I). In this case, every atom in the position x, y, z has an equivalent with the parameters $\frac{1}{2}+x, \frac{1}{2}+y, \frac{1}{2}+z$. If the first atom is moved to the origin of the cell, these positions are simplified to 0, 0, 0 and $\frac{1}{2}, \frac{1}{2}, \frac{1}{2}$. Now, the structure factor, simplified for the centrosymmetry, for an arbitrary reflection hkl may be calculated as the sum of these two atom contributions:

$$\begin{aligned} F_c(hkl) &= f\cos[2\pi(h\cdot 0+k\cdot 0+l\cdot 0)]+f\cos[2\pi(h/2+k/2+l/2)] \\ &= f+f\cos[\pi(h+k+l)] \end{aligned}$$

Since the sum $h+k+l$ must add to an integer, the cosine will have the value of $+1$ if it is even, and -1 if it is odd. Thus, unless $h+k+l = 2n$, $F_c = 0$: all reflections which fail to meet this *condition for reflection* are absent. Thus, for an I-centered lattice, half of the possible reflections are absent. In the sketch (Fig. 6.12) this is illustrated for the reflections 100 and 200.

An incident beam which has the right scattering angle for constructive interference by the 100 planes thus encounters an equivalent plane with half the spacing, and the path difference of $\lambda/2$ results in the cancellation of the intensity. On the other

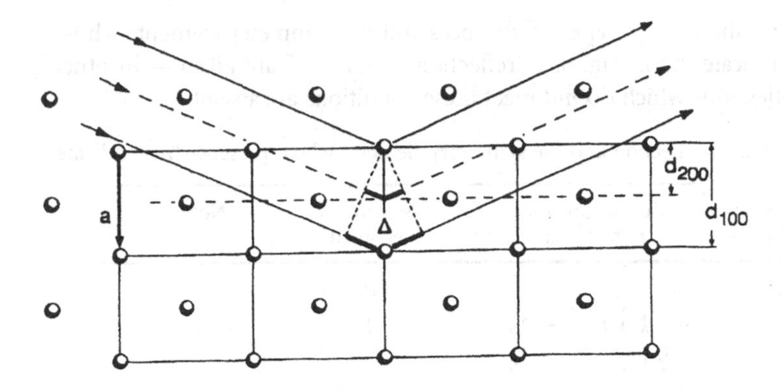

Fig. 6.12. Cancelling of intensity for the reflection 100 with I-centering.

hand, radiation encountering the crystal at the higher angle ϑ for the 200 reflection, for which there are (artificial) lattice planes with a spacing of $d/2$, is not cancelled by the interleaving planes, as they now scatter in phase, leading to an output intensity twice of what would have been given by the same planes in a primitive lattice.

All other symmetry elements with translation components lead to systematic absences in specific groups of reflections. Thus, the X-rays may be said to "see" the translation symmetry. From the domain of reflections which are affected by absences, the type of symmetry element and its orientation may be deduced. *General absences,* which affect a fraction of all reflections hkl indicate the presence of a centered lattice, and give directly the Bravais type. *Zonal absences,* which affect only planes in the reciprocal lattice: $0kl$, $h0l$, $hk0$ or hhl (zones are groups of lattice planes with one axis in common) indicate glide planes normal to $\boldsymbol{a}, \boldsymbol{b}, \boldsymbol{c}$ or [110] (see example in Fig. 6.13). *Row absences* affect only the reciprocal lattice lines: $h00$, $0k0$, $00l$ or $hh0$ and indicate screw axes parallel to $\boldsymbol{a}, \boldsymbol{b}, \boldsymbol{c}$ or [110].

The type of absence shows which type of centering is present, which glide direction a glide plane has, or which type of screw axis is present. Table 6.4 gives a

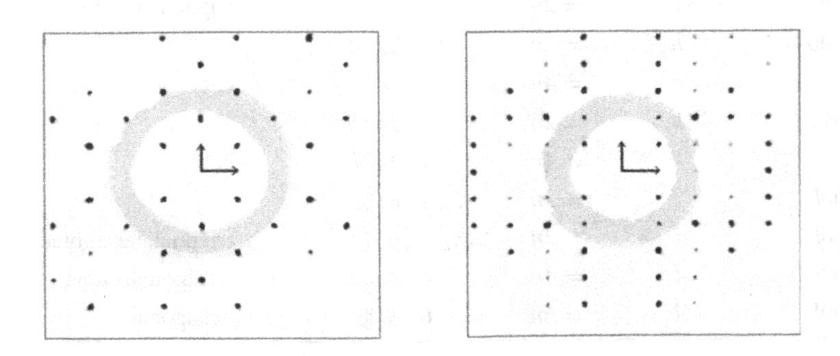

Fig. 6.13. Reciprocal lattice planes $h0l$ (left) and $h1l$ (right) (a^* vertical, c^* horizontal) with absences for an n glide $\perp$ $\boldsymbol{b}$ ($h0l$: $h + l \neq 2n$) and a b glide $\perp$ $\boldsymbol{c}$ ($hk0$: $k \neq 2n$).

summary of all of the various types of absences and the symmetry elements whose presence they indicate. As in Fig. 6.9, "reflection conditions" are given – in other words, those reflections which do *not* meet these conditions are absent.

Table 6.4. Conditions for reflection and the symmetry elements whose presence they indicate.

Domain of condition	Reflections affected	Conditions for reflections		Symmetry element	Notes
general	hkl	–		P	
		$h+k+l$	$=2n$	I	
		$h+k$	$=2n$	C	
		$k+l$	$=2n$	A	
		$h+l$	$=2n$	B	
		$-h+k+l$	$=3n$	R (obverse)	s. Sect. 2.2.1
		$h-k+l$	$=3n$	R (reverse)	
zonal	$0kl$	k	$=2n$	$b\perp a$	
		l	$=2n$	$c\perp a$	
		$k+l$	$=2n$	$n\perp a$	
		$k+l$	$=4n$	$d\perp a$	only F
	$h0l$	l	$=2n$	$c\perp b$	
		$h+l$	$=2n$	$n\perp b$	
		$h+l$	$=4n$	$d\perp b$	only F
	$hk0$	h	$=2n$	$a\perp c$	
		k	$=2n$	$b\perp c$	
		$h+k$	$=2n$	$n\perp c$	
		$h+k$	$=4n$	$d\perp c$	
	hhl	l	$=2n$	$c\perp[110]$	tetragonal, cubic
				$c\perp[120]$	trigonal
		$2h+l$	$=4n$	$d\perp[110]$	tetragonal, cubic I
	hhl	l	$=2n$	$c\perp a$	trigonal, hexagonal
row	$h00$	h	$=2n$	$2_1 \parallel a$	
		h	$=4n$	$4_1, 4_3 \parallel a$	cubic
	$0k0$	k	$=2n$	$2_1 \parallel b$	
		k	$=4n$	$4_1, 4_3 \parallel b$	cubic
	$00l$	l	$=2n$	$2_1, 4_2, 6_3 \parallel c$	
	$00l$	l	$=3n$	$3_1, 3_2, 6_2, 6_4 \parallel c$	trigonal, hexagonal
	$00l$	l	$=4n$	$4_1, 4_3 \parallel c$	tetragonal, cubic
	$00l$	l	$=6n$	$6_1, 6_5 \parallel c$	hexagonal

Diffraction symbols. The determination of the Laue group of a crystal reduces the possible crystal classes but still may leave a very large number of space group possibilities. The knowledge gained from the systematic absences in the diffraction pattern normally reduces the number of choices greatly. This information can be summarized in the diffraction symbol, proposed by Buerger, which gives first the Laue group, then the Bravais lattice and finally the symmetry elements whose presence has been established in the same order as they occur in the space group symbol. When an element in the symbol remains undetermined, it is represented by a dash. The only possible space groups are thus those which contain these symmetry elements with translation components. The diffraction symbol $2/mC-/c$, for example, indicates that the space group is monoclinic, C-centered, does *not* have a 2_1-axis parallel to $\boldsymbol{b}$, but does have a c-glide normal to it. It follows that the possible space groups are $C2/c$ and Cc.

The choice can be very easy, when a space group symbol contains *only* translational symmetry, e.g. in the case of the most commonly occurring of all space groups, $P2_1/c$ (Nº 14). The choice is much greater when, particularly in the orthorhombic system, a choice has to be made between 2-fold axes and mirror planes, as these elements are distinguished neither by Laue symmetry nor by the systematic absences. If an orthorhombic crystal has no systematic absences (diffraction symbol $mmmP---$), the space groups $P222$, $Pmm2$ and $Pmmm$ are all possible, and in addition, $Pmm2$ has three possible settings. The 2-fold axis can lie parallel to $\boldsymbol{a}$, $\boldsymbol{b}$, or $\boldsymbol{c}$, giving the space group symbol $P2mm$, $Pm2m$, or $Pmm2$. Once their orientation is known, the axes of the cell should be named so as to get the standard setting, $Pmm2$. For less experienced workers, International Tables A Table 3.2 gives a sort of plant identification key!

When two or more choices of space group remain, other criteria must be used. These may include physical properties such as piezoelectricity, Patterson symmetry (cf. Sect. 8.2) or chemical plausibility. If necessary, it is possible to attempt to solve and describe the structure in all possible space groups. Usually, only the correct one will give a satisfactory solution (cf. Ch. 11). The example of space group $Pmm2$ indicates a frequently occurring problem: the need to transform the original unit cell.

6.7 Transformations

The best way to describe transformations of unit cells is as a multiplication of a 3×3 matrix, the *transformation matrix* by the column matrix consisting of the old axes $\boldsymbol{a}$, $\boldsymbol{b}$, $\boldsymbol{c}$.

$$\begin{pmatrix} t_{11} & t_{12} & t_{13} \\ t_{21} & t_{22} & t_{23} \\ t_{31} & t_{32} & t_{33} \end{pmatrix} \begin{pmatrix} a \\ b \\ c \end{pmatrix} = \begin{pmatrix} a' \\ b' \\ c' \end{pmatrix} \tag{6.3}$$

This will give the new axes $\boldsymbol{a}'$, $\boldsymbol{b}'$, $\boldsymbol{c}'$ in terms of the old.

$$\begin{aligned} a' &= t_{11}a + t_{12}b + t_{13}c \\ b' &= t_{21}a + t_{22}b + t_{23}c \\ c' &= t_{31}a + t_{32}b + t_{33}c \end{aligned}$$

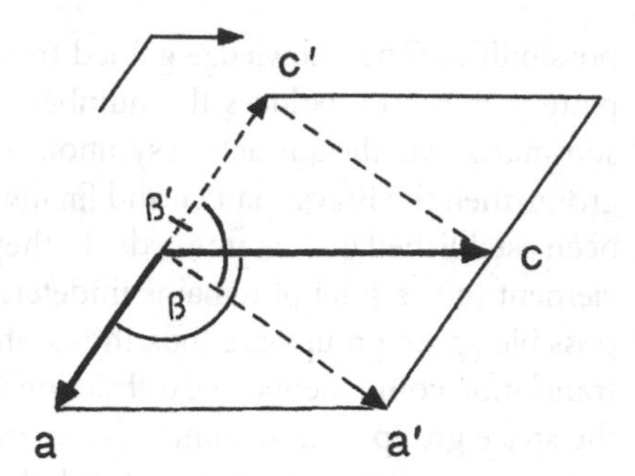

Fig. 6.14. Alternative settings of a monoclinic cell in space group $P2_1/c$ (solid) and $P2_1/n$ (dashed).

Since these are vector operations, the magnitudes of the new lattice constants and angles must be calculated explicitly. As a practical example, the transformation of a monoclinic cell in the space group $P2_1/c$ into one in the space group $P2_1/n$ will be given (Fig. 6.14). These are, in fact alternative settings of the same space group. As has already been suggested, the setting should be chosen which makes the monoclinic angle at least 90°, and as close as possible to that value, as this leads to more straightforward refinement and description of the structure (cf. Chap. 9 & 12). Since, in this transformation, the direction of the glide becomes a diagonal, instead of lying along **c**, the symbol in the space group is *n* instead of *c*.

$$\text{transformation matrix:} \quad \begin{pmatrix} 1 & 0 & 1 \\ 0 & 1 & 0 \\ \bar{1} & 0 & 0 \end{pmatrix}$$

$$a' = \sqrt{a^2 + c^2 - 2ac\cos(180^\circ - \beta)}$$

$$c' = a$$

$$\beta' = 180^\circ - \arccos\frac{a^2 + a'^2 - c^2}{2aa'}$$

Depending on the shape of the unit cell, it is possible that the choice of the alternative face diagonal is preferable. This is often easy to see by making a simple sketch. A similar situation arises in making a choice between the alternative space group settings $C2/c$ and $I2/a$. Despite that fact that $I2/a$, like $P2_1/n$ got into discredit as an unconventional setting, (neither of them are set out in the older International Tables I) it is very reasonable to use them when they give a monoclinic angle closer to 90°. In any case, it is never sensible to work with angles > 120°.

An interesting case arises with the cubic space group $Pa\bar{3}$ (№ 205, full symbol $P2_1/a\bar{3}$, which is a cubic supergroup of *Pbca* (№ 61): In spite of the cubic symmetry, it is not possible to interchange the axes arbitrarily, as that may change the direction of the glide. The symbol *a* indicates that there is an *a*-glide perpendicular to **c**, and hence a *b*-glide perpendicular to **a**, and a *c*-glide perpendicular to **b** as in *Pbca*. This will be shown by the systematic absences. The alternative setting, analogous to *Pcab*, would be $Pb\bar{3}$, with different absences.

CHAPTER 7

Experimental Methods

This chapter deals with the most important methods for obtaining the necessary data for an X-ray single crystal structure determination. The first step, naturally, is the obtaining of a suitable crystal.

7.1 Growth, Choice and Mounting of a Single Crystal

The development of a crystal depends on the relative rates of nucleation and growth. If the rate of nucleation is larger than the rate of growth, the result will be agglomerates of small crystallites. On the other hand, too rapid a rate of growth may result in the inclusion of many faults in the crystal. The way to avoid both of these problems, is, unfortunately, something which cannot readily be predicted at the beginning of the study of a new compound, and often requires "green fingers". Crystallisation may be attempted from solution, from a melt, or by sublimation. A few simple examples and practical suggestions will be given here.

Crystal growth from solution. The substance to be crystallized should be only moderately soluble. The most usual procedure is the slow cooling of a saturated solution, often by placing the container in a polystyrene jacket or a Dewar flask in order to allow it to cool very slowly. In order to reduce the number of nucleation sites, new, smooth glass or PTFE vessels are best. The set-up should be allowed to stand in a vibration-free environment, e.g. not near a working pump. When possible, higher temperatures are preferable to lower (e.g. in a refrigerator or deep-freeze) as they minimize the chance of inclusion of unwanted solvent of crystallization. Often, arranging a small temperature gradient is useful, perhaps by placing the glass vessel in a sloping hole in a slowly cooled metal block so that the upper part of the solution remains outside. Convection will then transport material through the solution.

In cases of difficulty, different solvents should be tried, avoiding, if possible those like $CHCl_3$ and CCl_4, which contain heavy atoms and are often included in the structure disordered so that they detract from the accuracy of the structure determination.

If the compound is synthesized by the mixing of two reagents in different solvents, in favorable cases, direct crystallization will occur at the solvent interface. Good crystals are often obtained when the solvents diffuse slowly into one another.

In more difficult cases, the rate of diffusion can be further reduced by the use of a dialysis membrane or a gel. In a variant of the diffusion method, a second solvent, in which the solute is less soluble is allowed to diffuse into a solution, either through a membrane or via the gas phase *(isothermal distillation).*

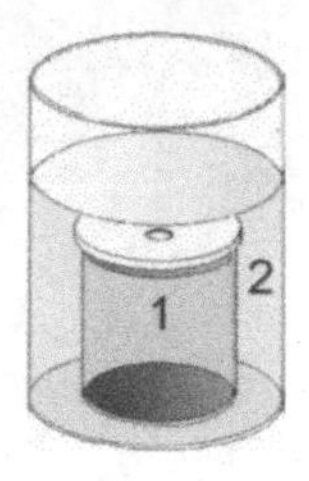

Fig. 7.1. shows a variant of this *diffusion* method, which is very simple to devise but often very successful. A small container is filled with solution 1, covered by a lid with a small hole, and carefully lowered into a large container of solution 2.

Hydrothermal methods. These methods are particularly applicable to very slightly soluble inorganic materials. They are put into a small autoclave with an aqueous solvent (water, alkali, HF, etc.) in which they are normally insoluble. Heating the mixture, usually to 200–600°, so that the pressure reaches several hundred atmospheres, results in a supercritical fluid in which many compounds will dissolve and recrystallize on slow cooling. Hydrothermal methods are often useful for the synthesis of such compounds. Since ordering is favored at high pressures, good crystals are often obtained in a relatively short time.

From a melt, crystals suitable for X-ray analysis are seldom obtained. This is because the chunks of solid obtained from the melt must be cut in order to obtain single crystals of suitable size. Thus, the crystals rarely have well-defined faces, making it difficult to orientate them for film methods or to measure them for numerical absorption correction (cf. sect. 7.4.3). The crystal quality is also often worse after this treatment. For some inorganic materials, however, this is the only possible method of obtaining crystals. In a modification of this method, the melt is sintered about 10° under the melting point, and crystals of suitable size sometimes develop either on the surface or in the body of the partially liquid mass. For this method, it is useful to undertake DTA- or DSC-studies of the material in order to find the best temperature range. They will also indicate any phase changes, passing through which will be likely to give twinned crystals.

Sublimation. Sublimation can give very good crystals, but has been little used. The same is true of the method of *chemical transport.* If a suitable transport medium can be found, many chalcogenides and halides — particularly binary ones — can be grown as beautiful crystals — in fact often too big for X-ray examination. For further information on crystallization, the review article by Hulliger [23] is recommended.

Crystal size is an important consideration, and the ideal value depends on the chosen radiation and the accompanying absorption. In any case, dimensions greater than 0.5 mm should be avoided, as the uniform region of the X-ray beam is usually no bigger than this. For weakly diffracting crystals, particularly very thin needles, a length of up to 0.8 mm is all right, since the gain in intensity more than offsets the errors due to inhomogeneity of the beam. In such cases, the softer Cu-radiation (if available) is recommended because the interaction with the electrons is approximately proportional to λ^3. Thus, the soft Cu-radiation gives up to 8 times higher intensities with respect to the harder Mo-radiation. Since, however, the absorption factor for Cu-radiation is, on the average, about 10 times that for Mo-radiation, the use of Cu is largely limited to light atom structures or very small crystals of compounds with heavier elements. For crystals giving strong absorption, an optimum size can be estimated, as the scattering capacity increases directly with the crystal volume, while the absorption increases exponentially with the crystal thickness. As a rule of thumb the ideal mean thickness will be $2/\mu$, where μ is the absorption coefficient for the given wavelength. More about the problem of absorption is given in section 7.4.3.

Crystal quality is best judged using a polarising stereomicroscope with a 20- to 80-fold magnification. Most crystals are transparent. Only cubic crystals are optically isotropic, i. e. have an index of refraction independent of direction. All other crystals are *optically anisotropic*. Tetragonal, trigonal and hexagonal crystals are said to be *optically uniaxial*. This means that that their index of refraction along the c-axis is different from that in the a, b-plane. Orthorhombic, monoclinic and triclinic crystals are *optically biaxial*, as they have different refractive indices in all three directions. All optically anisotropic crystals rotate the plane of polarized light, except for uniaxial crystals viewed directly along the c-axis. In the polarising microscope, the crystal is viewed by polarized transmitted light. A second, rotatable polarization filter is placed in front of the objective, and rotated to give a dark field (crossed polarizers). When a crystal rotates the plane of polarization, it appears light against the dark background. As the crystal is rotated, a position will be reached where the crystal goes dark ("extinguishes") and then becomes light again on further rotation. This extinction occurs every 90°. A multiple crystal, that is one made up of two or more crystallites, the individual parts will normally extinguish at different angles. Cracks in a crystal are often revealed as bright lines against the dark background. All such crystals should be rejected, or possibly divided carefully with a scalpel, so that all unwanted regions are removed. In some cases, the crystal must in any case be cut to obtain the desired shape and size. This operation is best carried out under a drop of inert oil, e. g. silicone oil, on the microscope slide, to prevent the cut pieces from springing away. Crystals that are too large can sometimes be reduced in size on the microscope slide by the careful application of a drop of solvent. Crystals that are chemically and mechanically stable can be freed of unwanted "passengers" by rubbing them between fingers, coated with a small amount of grease.

Crystal mounting. Crystals are then usually mounted on a thin glass fiber, using some adhesive such as grease, shellac, a two-component adhesive or "super glue". This cheap and convenient method has the disadvantage that the glass fiber, which must be robust enough to prevent movement of the crystal in the beam, will itself contribute to absorption and background scatter of X-rays. Alternatively, using a glass fiber, the crystals can be mounted in a glass or quartz capillary with a diameter of 0.1–0.7 mm and a wall thickness of 0.01 mm. To hold the crystal still during data collection, some grease is usually added as well. This is usually best done by first inserting the grease with a fine glass fiber from one end of the capillary, and then sealing it. The crystal is then introduced from the other end, which is then also sealed to protect the crystal. It is also possible to mount a crystal by attaching it to the outside of such a capillary, as it is stiffer than a fiber while placing less glass in the X-ray beam.

The capillary or the fiber is then attached with adhesive to a small metal or fiber rod, suitable for mounting on a *goniometer head*. Goniometer heads are manufactured to standard sizes and mountings, and are suitable for nearly all cameras and diffractometers. They all have two mutually perpendicular slides, which allow the crystal to be *centered* accurately on the axis of the head. For film methods, it is also necessary to have two mutually perpendicular arcs, to allow the crystal to be *oriented* (Fig. 7.2). It is also useful to have a height adjustment for use on single crystal diffractometers which generally have no such adjustment on themselves. Because the adjustment range is limited to a few mm, it is best to have the glass fiber or capillary ready for mounting so that the crystal will be more or less in the "correct" place before any adjustment is made. A two-circle optical goniometer is a useful tool both for mounting and for measuring the crystal, and it can be calibrated so that the crystal can be brought to the correct position.

It is more difficult to mount *air sensitive crystals*. There are three main choices of method. Selection of the crystal and mounting in a capillary can be carried out on a microscope in a glove box. Instead of sealing the capillary with a flame, it can be closed with a drop of super glue. The second method is to use a special type of Schlenk tube, with fitted capillaries. The crystal is inserted with the help of a long rod with a bent tip and sealed in a stream of argon. The simplest method, and the one most likely to prove successful with very sensitive crystals is to remove the crystals from a Schlenk line in a stream of argon into a drop of dry, inert oil on a microscope slide. This can then be placed on the stage of a polarizing microscope, where the crystal can be inspected and, if necessary, cut. With the help of a little grease the crystal is then attached to a precentered glass fiber, and quickly mounted on the diffractometer in the path of cold nitrogen. At temperatures of less than −80°C, there is normally no danger of further decomposition during the time taken to measure the diffraction data. The main disadvantage of this method is that the crystal shape can no longer be measured accurately, so no exact absorption corrections can be made.

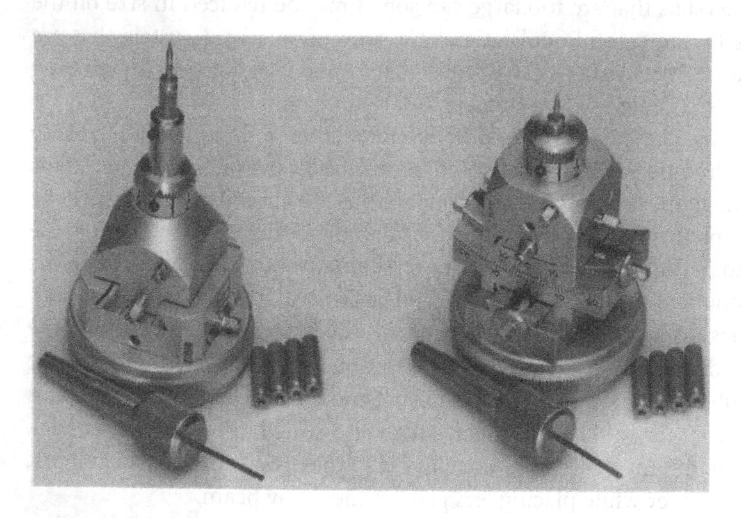

Fig. 7.2. *XY*-Goniometer head for diffractometers and area detectors (left) and with additional arcs for film recording (right). (By permission of Huber.)

7.2 Measuring the Diffraction Pattern of Single Crystals

Once a crystal has been mounted, the actual measurement of the diffraction pattern can begin. This involves the measurement of a large number of reflection intensities, normally between 1000 and 50 000. Since each reflection requires a particular orientation of the corresponding lattice planes (hkl), the crystal must be readjusted relative to the X-ray beam in order to bring each set of planes into the "reflecting position". This is equivalent to saying that the reciprocal lattice (Chapter 4) must be rotated to bring each scattering vector $\boldsymbol{d}^*(hkl)$ into the position where it cuts the Ewald sphere.

In recent years, the techniques for achieving this have developed dramatically. Until the early 1970's, intensity data were mainly collected using cameras of various sorts. The next twenty years were the era of the four-circle diffractometer, and these have now largely been replaced by area-detector systems. In this text, only a brief introduction is given to classical film methods and the most important types of four circle diffractometers. The main emphasis is on work using area detectors. Since these combine speed and accuracy with the advantages of film methods, they are useful both for routine measurement and for teaching basic crystallographic principles.

7.2.1 Film Methods

*The Laue Method:*The oldest film technique uses a stationary crystal and unfiltered "white" radiation. The result, in terms of the Ewald construction, is the superposition of many Ewald spheres, the largest of which has a radius corresponding to the shortest wavelength in the beam (see Section 3.3, equation 3.2). All of the reciprocal lattice points lying inside this sphere are in the diffracting position and are projected onto the plane of the film. They each record with individual wavelengths and crystal-film distances, so the pattern is not simple to interpret! They can, however, show the symmetry of the crystal viewed from the direction of the X-ray beam.

Rotation and Weissenberg Methods: For the explanation of this technique, the Ewald construction given in Fig. 4.2 is very useful. The crystal, on its goniometer head, is mounted about a horizontal rotation axis, at 90° to the X-ray beam. A metal cylinder is placed over this, containing a slit to allow the X-ray beam to enter. On the inside of this, protected by a piece of black paper, is the X-ray film (Fig. 7.3).

If the crystal is adjusted so that a real axis is parallel to the rotation axis, i. e. reciprocal lattice layers are perpendicular to it, these layers will record as "layer lines"

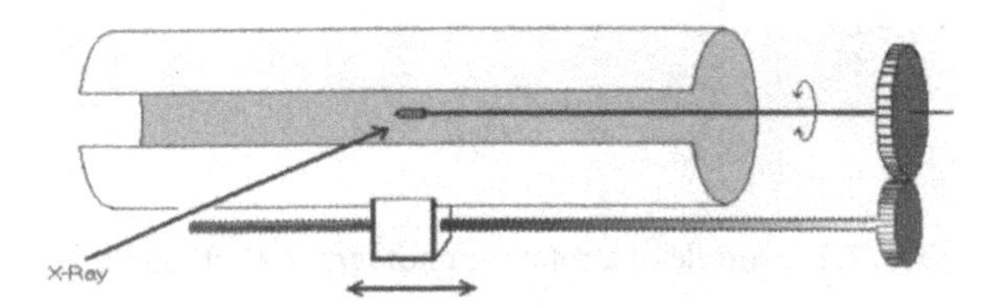

Fig. 7.3. Diagram of a Weissenberg Camera.

on the film, giving a "rotation photograph" (Fig. 7.4 left). If screens are fitted so that only one of these lines can record, and the film is moved parallel to the axis of the cylinder as the crystal rotates, the resulting picture (a Weissenberg photograph) is a distorted picture of that reciprocal lattice layer (Fig. 7.4 right).

Precession method: The precession camera (also called a Buerger camera after its inventor) works by a different principle, but one which is also easily understood in terms of the Ewald construction. In this case, a reciprocal lattice layer is adjusted to be *normal* to the X-ray beam (Fig. 7.5). Instead of rotating the crystal and the chosen reciprocal plane, it is tilted with respect to the Ewald sphere by a constant angle, μ, the precession angle. Thus, the plane cuts the sphere in a circle with a radius dependent on μ. Those reflections which happen to lie on this circle will be recorded. If the crystal is now allowed to precess, the normal to the reciprocal plane will describe a cone with a cone-angle of 2μ. The circle of intersection thus moves about the origin of the plane, and brings all reflections into the diffracting position at some time on the circumference of a circle with twice the radius of the circle of intersection. A flat film is then made to precess along with the crystal in such a way that the crystal to film distance is always the same for any reflection when it intersects the Ewald sphere, and so records an undistorted projection of the reciprocal plane (Fig. 7.6). A screen, containing a ring-shaped slit, precesses along with the film to prevent more than one reciprocal layer recording.

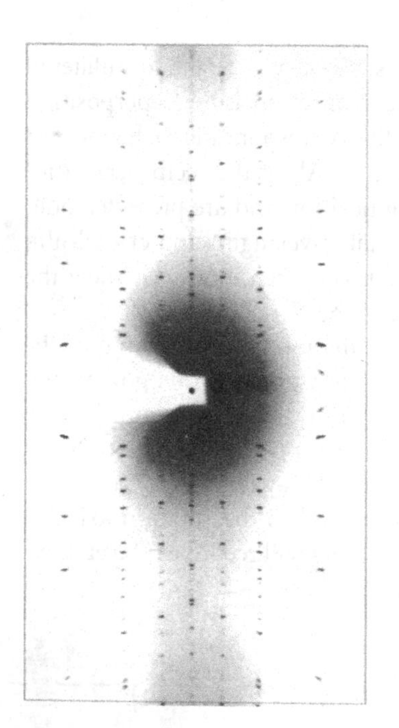

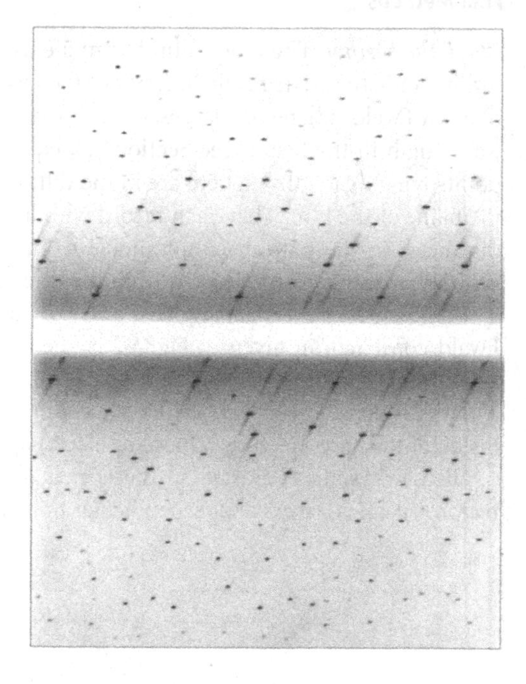

Fig. 7.4. Example of a rotation photograph (left) and a Weissenberg photograph (right).

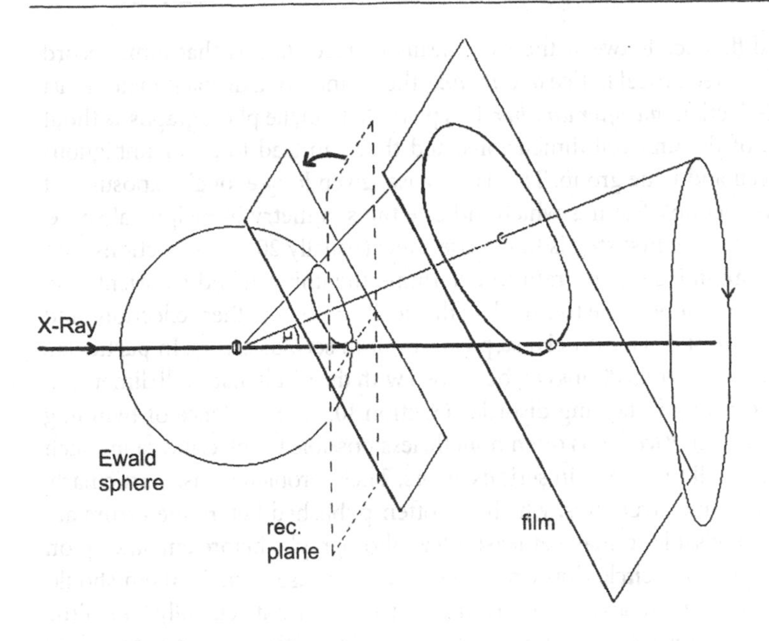

Fig. 7.5. Diagram of a precession camera.

Weissenberg and precession photographs on a well-aligned crystal are invaluable for the study of a wide range of diffraction phenomena. Consequently, they are still valuable tools, when a four-circle diffractometer is used for intensity measurements.

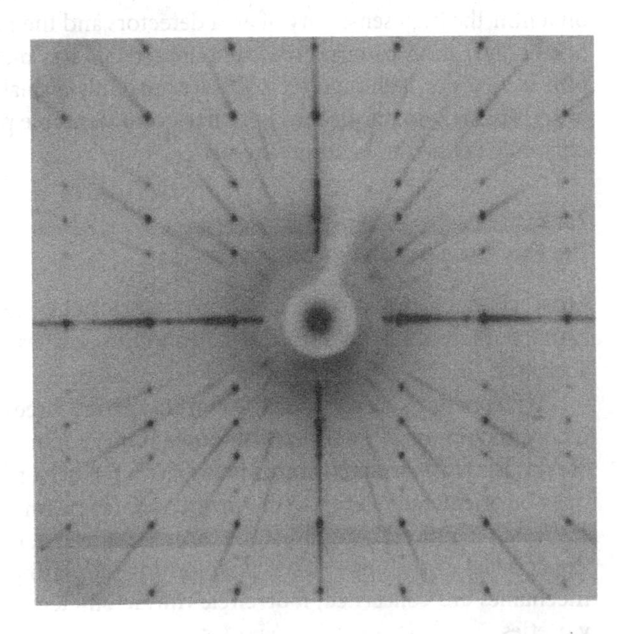

Fig. 7.6. Example of a 0-layer precession photograph.

The principal difference between the two means of recording is that films record entire layers of the reciprocal lattice *at one time*, the counter of a diffractometer must measure each reflection *one after another*. It is possible to make photographs without any knowledge of the unit cell dimensions, and they can lead to an unambiguous choice of unit cell and Laue group. This is because, given long enough exposures, it is impossible to overlook features which indicate the symmetry in reciprocal space. On a diffractometer, the first step is to choose a few (usually 20–40) reflections, and with the help of an indexing program to determine the unit cell and its orientation in space. Only then is it possible to calculate the positions of all other reflections and measure their intensities. This crucial step can result in serious errors. In particular, weak ("superstructure") reflections can be missed with the result that a cell dimension is halved, and streaking indicating disorder (Section 10.1), or evidence of twinning (Section 11.2) goes unnoticed. It is often nonetheless possible to solve and refine such a structure, which will then contain serious errors. These problems arise particularly in solid-state inorganic structures, which are often published before the errors are found. It is thus advisable to make at least a few photographs before embarking on data collection on a four-circle diffractometer, or in any case to make them should the solution and refinement give any indication of irregularity. It generally takes little time to make a few photographs compared with the time squandered on struggling with a wrong unit cell or space group!

Making photographs is, however, very time- and labor-intensive, since in addition to the time taken to adjust the crystal, exposures of 10–50 hours will be required for each photograph, and a complete set will normally require up to three weeks. Similar information is now available with area-detector systems, and requires less than a day. While it is true that individual reflections appear less sharp on an area detector than on a film, the high sensitivity of area detectors and their ability to display any section of the diffraction pattern on a computer screen has meant that they have displaced film techniques. Although the software currently available for the various types and brands of area detectors does not offer every desirable possibility, it is clear that they currently define the "state-of-the-art".

7.2.2 The Four-circle (serial) Diffractometer

Most laboratories have working four-circle diffractometers, and they are valued because of their ability to measure reflection intensities automatically and with high accuracy.

The apparatus currently on the market all have three computer controlled "circles" whose rotation axes intersect with one another to an accuracy of about 10 μm. The crystal has to be centered at this point. With the use of servo motors, the crystal can then be orientated relative to the incident X-ray beam so that the Bragg condition is met *and* the reflection occurs in the horizontal plane, in which a fourth circle brings the counter to the appropriate position for measuring the reflection. As far as the mechanics are concerned, four-circle diffractometers are made in one of two main varieties.

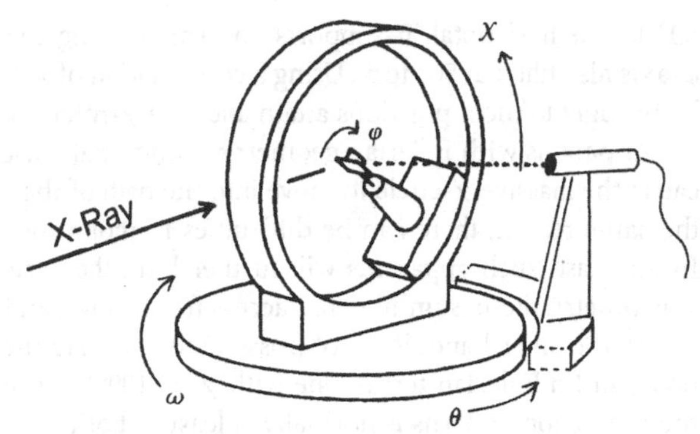

Fig. 7.7. Four-circle Eulerian-cradle diffractometer.

Eulerian geometry: The goniometer is mounted firstly on an ω-circle, lying in the horizontal plane. Perpendicular to this is the vertical χ-circle, and the goniometer head is mounted so that it can travel inside this. The goniometer head itself is mounted on and coaxial with the third or φ-circle. Finally, the fourth or ϑ-circle, which is coaxial with ω, carries the detector (Fig. 7.7). The detector may be a proportional or scintillation counter. Diffractometers operating in this way are manufactured by Huber, Rigaku, Stoë and Bruker-AXS (formerly Bruker, formerly Siemens, formerly Nicolet, formerly Syntex, formerly Scintag), and previously by Hilger and Watts, Philips and Picker.

Kappa geometry: Another way to orientate the crystal in space is that of the MACH3-, formerly the CAD-4 Diffractometer, made by AXS-Nonius (formerly Enraf-Nonius, Fig. 7.8) and the analogous device of CUMA, now sold by Oxford Instruments as "Xcalibur". The ω- and ϑ-circles are identical to those of the Eulerian instruments. The χ-circle is replaced by the "$\varkappa$-circle", the axis of which is tilted at

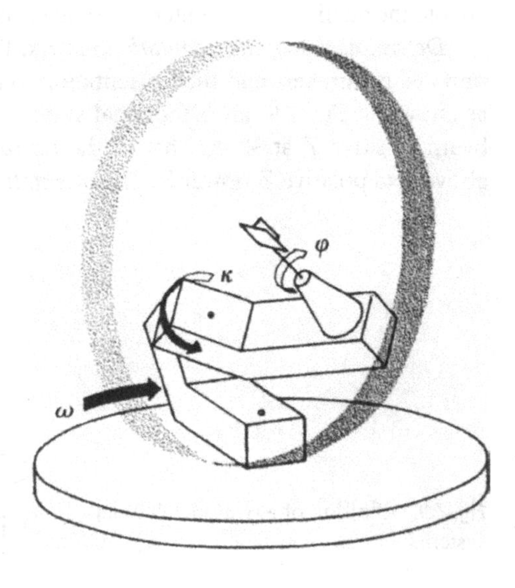

Fig. 7.8. The CAD4 $\varkappa$-axis diffractometer (by permission of Enraf-Nonius).

50° to the horizontal. It supports an arm carrying the goniometer head, with the φ-axis also tilted at 50° to κ. Using a combination of κ and φ settings, the crystal can be brought to most positions attainable by a χ-rotation in Eulerian geometry.

Apparatus with Eulerian geometry cannot make use of those values of ω which cause the massive χ-circle to move into the path of the direct or scattered beam. For the same reason, there can be difficulties in mounting a device to cool the crystal. In any case, such apparatus will further limit the accessible range of angles. With κ-geometry, there is unhindered access from above, and there are no restrictions on ω. On the other hand, it is not possible to orientate the goniometer in a "hanging" position (in Eulerian terms, one with $\chi > 100°$). Since, however, the symmetry of the diffraction patterns is normally at least $\bar{1}$, both types of mounting can normally measure a full set of "independent" reflections. For most diffractometers, the X-rays are monochromatized using a graphite crystal.

Crystal centering: On a four-circle diffractometer, a crystal need not be orientated, only centered accurately. To achieve this, the φ-axis is orientated perpendicular to the axis of a microscope mounted on the diffractometer so that these two axes define a vertical plane. By rotation of the φ-axis, one of the goniometer head slides is set horizontal, and the position of the crystal relative to the vertical cross-wire of the microscope is noted. The crystal is then rotated through 180°, and the measurement repeated. The true center, where the crystal is rotated, is the average of the two readings. The operation is then repeated for the other slide after a 90° rotation of φ. To adjust the height on an Eulerian instrument, the position of the crystal is observed at $\chi = +90°$ and $-90°$. On an instrument with κ-geometry, two settings are found with the φ-axis horizontal and perpendicular to the microscope axis. In either case, the height is adjusted to center the crystal fully. The accuracy of both the unit cell parameters and the subsequent intensity measurement are dependent on a stable and accurate centering of the crystal. The software provided with the instrument will enable the quality of the centering to be checked.

Determination of the orientation matrix. The next step is the determination of the unit cell parameters and their orientation to the goniometer axes. These are defined as shown in Fig. 7.9: an orthogonal system, with positive X pointing at the X-ray beam, positive Y at 90° to that in the horizontal plane, clockwise as viewed from above, and positive Z upwards. The *orientation matrix* is a 3 × 3 matrix which gives

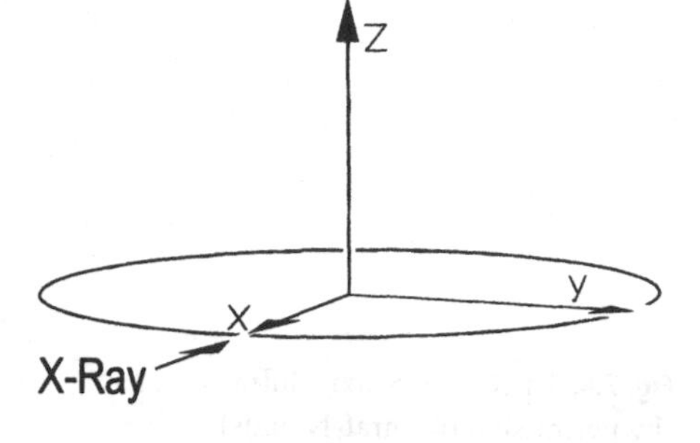

Fig. 7.9. Definition of a goniometer axial system.

(in $Å^{-1}$) the components of the three reciprocal axes in the three directions of the goniometer's axial system. It therefore contains the basic data for the definition of the reciprocal unit cell and its orientation in space. Once it is known, the position of each reciprocal lattice point may readily be calculated. *(The definition of the orientation matrix does vary somewhat from one instrument to another, and reference should be made to the manufacturer's instructions.)*

The determination of the orientation matrix and thereby the unit cell parameters may be accomplished in three ways. In every case, a selection of reference reflections is located, spread as widely as possible in reciprocal space. Once the angles corresponding to these positions have been optimized, the matrix can be refined by the method of least squares. The following strategies are all possible.

1. If the crystal had previously been orientated and its unit cell determined by film methods, it is possible to make use of the relationship between the axial systems of the camera and the diffractometer. The positions of a few strong reflections may be estimated, and these positions found and refined. The refined positions, along with their *hkl* indices may then be used to define the matrix.
2. Even without the crystal having been orientated, a flat polaroid film cassette can be used to take a rotation photograph, and the film coordinates for a few reflections used to give rough diffractometer angles. These are then optimized to give a set of reciprocal lattice points, and their positions are calculated relative to the goniometer axial system. The reciprocal cell parameters remain unknown, so an indexing program is then invoked which will apply one of several strategies to select a reciprocal cell, which will define a lattice onto which all of the observed reflections will fit. This will then give the positions of the reciprocal lattice basis vectors and hence the orientation matrix.
3. Lastly, the method probably most commonly used is to make a random search for reflections, which are then optimized and used for the same indexing procedures given above.

This step in a structure determination is critical, and must be critically inspected when no film data are available. Even when films have been taken, the automatic search is often used because of its great convenience. In such cases, the films should be rechecked to make certain that the same unit cell has been found. An advantage of this dual approach is that it can also correct errors in film interpretation! For the proper refinement of the orientation matrix, it is important to have sufficient strong reflections well distributed in the reciprocal lattice. For this purpose, photographs are an advantage for selecting further strong reflections; otherwise, these must be sought, using the initial, approximate matrix. The less symmetric the unit cell, the more cell parameters require to be refined, and the more important is the number of reference reflections, which should in any case be at least 20.

The assignment of the "correct" Bravais lattice is now made, by analysis of the reduced cell. It must not be forgotten that metric considerations alone do not define a crystal class.

Problems with indexing. When crystals have not been photographed, it sometimes happens that not all of the selected reflections can be indexed. This may be because the indexing program has not yet locate the correct cell, and may be cured by locating other data in a different part of reciprocal space. The more common reason is that the crystal is not truly single, and that some of the reflections are in fact those of a "passenger" or possibly another twin component. In such cases, two or more orientations of the reciprocal lattice are superimposed on one another (cf. sect. 11.2). If there are too many of these "foreign" reflections in the list, it may not be possible to index it at all. Sometimes the omission of, say, all but the strongest data will help. Programs such as DIRAX [77] can help to reject those data not indexing on the most consistent cell. If only a few foreign reflections affect the data, it is likely that the structure can be solved successfully. In worse situations, it may be impossible to solve the structure without determining the twin law, perhaps by film methods. In such cases, it is often best to look for another crystal!

7.2.3 Reflection profile and scan type

The usual way to measure intensities on a four-circle diffractometer is to move the four circles to the calculated positions for the reflection. Then one of the circles, usually ω, is moved to a calculated position away from the maximum (e.g. 0.5°) then, with the radiation window open, that circle is moved slowly (e.g. 1–10° min^{-1}) through the diffracting position. In terms of the Ewald construction, this is equivalent to moving the scattering vector $\boldsymbol{d}^*$ slowly through the Ewald sphere by the scanning angle $\Delta\omega$. A plot of the measured intensity as a function of ω (Fig. 7.10) will show that the reflection has a specific "profile" above background. The full reflection width (background to background) for a good, small crystal is about 0.5–0.8°, but in badly

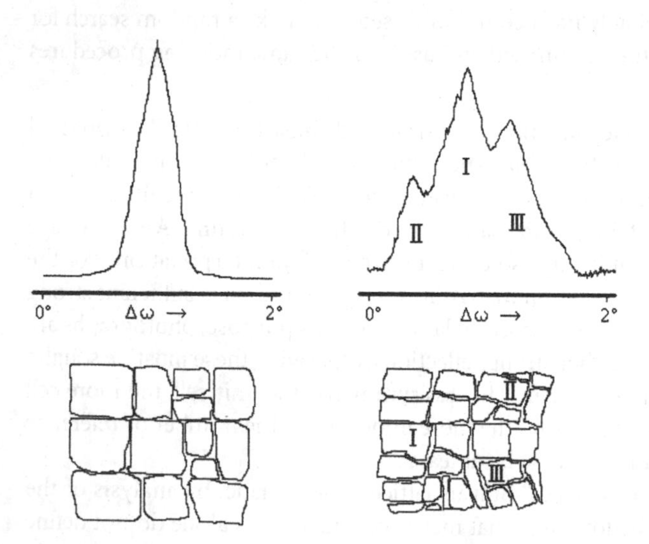

Fig. 7.10. Reflection profile and mosaic structure (greatly exaggerated).

crystallized or mechanically strained specimens, it may be as much as 2–3°. Crystals with such broad profiles usually show splitting and usually are not worth measuring.

Mosaic structure. Such broadening and splitting of reflection profiles are related to the *mosaic structure* of a crystal. Real crystals only show the ideal lattice structure over a relatively small volume. Beyond that, crystal faults cause the mosaic blocks to be misoriented with respect to one another by 0.1–0.2°. These small imperfections, in fact, simplify the measurement of intensity, and are the basis of the validity of calculated diffraction intensities (sect. 10.3).

Poorer crystals show a broadened profile, which leads to poor peak shapes and difficulties in determining the peak position. This reduces the accuracy of the cell parameters and of the orientation matrix. As may be seen in Fig. 7.11, reflection broadening as a result of coarse mosaic structure is very easy to detect in rotation photographs. In order to plan data collection, it is necessary to measure several profiles in order to determine an optimal scan width.

*Scan method.*Intensities can be measured using the pure ω-scan, which has just been mentioned, by holding the counter stationary at the calculated 2ϑ position, or alternatively by the $\omega/2\vartheta$ method, sometimes called the $\vartheta/2\vartheta$ method. In this case, the ω-circle is first positioned at ½$\Delta\omega$ below its calculated position, and the counter at $\Delta\omega$ below its calculated position. The reflection is then scanned with the circles moving simultaneously through $\Delta\omega$ and $2\Delta\omega$ respectively. The $\omega/2\vartheta$ scan is particularly appropriate when the radiation has been monochromatized using filters. It is also useful for measurements at very high scattering angle, normally with Cu radiation, when the divergence of the $K_{\alpha1}$ and $K_{\alpha2}$ lines must be taken into account. The method should *not* be used for crystals with large mosaic spread, as it gives apparently narrower reflection profiles with much larger errors in intensity. This happens because when very wide scans (e.g. $\Delta\omega = 2°$) are needed in order to include the scattering of all the mosaic blocks, the scan begins with the counter at $\Delta\omega$ less than the required 2ϑ value. In terms of Fig. 7.10, when the blocks in region II are in the diffracting position, the scattered radiation does not reach the counter. The mosaic blocks in region I give rise to a relatively sharp reflection, since the counter is correctly

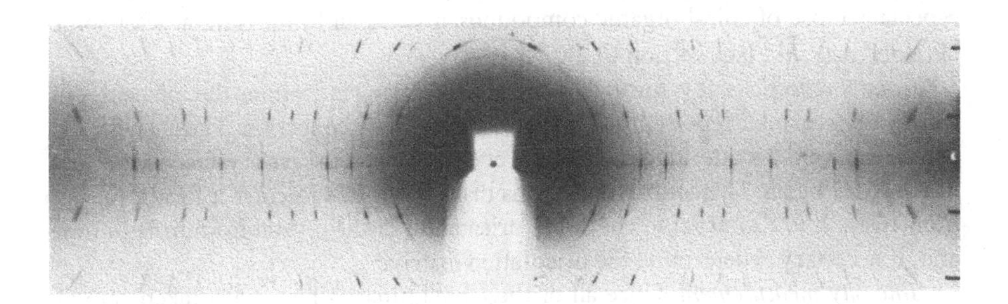

Fig. 7.11. Oscillation photograph of a crystal with a very substantial mosaic structure, so that it shows partial Debye-Scherer rings characteristic of a powder photograph.

positioned for them. The intensity from blocks in region III, which occurs at the end of the scan is again lost through the bad positioning of the counter. It is thus always best to use reasonably broad, pure ω-scans when a graphite monochromator is in use and the scan profiles have not been fully characterized (Fig. 7.10, upper right).

Selection of Data to be Measured. In addition to the decision of how reflections are measured, a choice must be made of exactly what portion of reciprocal space is to be covered. The chosen maximum value of the scattering angle ϑ is one limit. It defines the sphere in reciprocal space outside of which data are not measured. With Mo radiation, this should be at least 25°, in no case less than 22°; the corresponding minimum lattice plane spacing, referred to as the "resolution" of the data is here 0.84 or 0.95 Å, respectively. For copper radiation, the same resolution corresponds to a maximum ϑ of 66° or 54°. The greater the scattering power of the crystal, i.e. in particular when there are heavy atoms present, the higher is the limiting value of ϑ – sometimes as much as 40° and 75° respectively. Normally, there is also a lower limit of 2–3°, since at very low angles, the proximity of the direct beam makes the measurements inaccurate.

A decision must also be made as to the segment of the sphere in reciprocal space that is to be measured. This is dependent on the Laue symmetry. For triclinic crystals, at least a hemisphere must be measured, for monoclinic at least a quadrant: the Laue group $2/m$ implies that there is a mirror plane in reciprocal space normal to the $\boldsymbol{b}^*$ axis, resulting in the equivalence of the hkl and $h\bar{k}l$ reflections. At the same time, the 2-fold axis parallel to $\boldsymbol{b}^*$ makes hkl equivalent to $\bar{h}k\bar{l}$, so it is only necessary to measure data with all indices positive and those with either h or l negative. For orthorhombic crystals, only an octant (hkl all zero or positive) need be measured, since the mmm symmetry makes the other seven possibilities equivalent. In the trigonal system, Laue group $\bar{3}$, the minimum data required are those of the type $\bar{h}kl$ (h only negative and the other indices only positive). This is, in fact, two thirds of a hemisphere, since the reflections hkl (all positive) only cover a 60° sector of the hemisphere. For higher symmetries still less need be measured, but, especially in cases of doubt, it is always best to measure more than the minimum required if measuring time is available. With duplicate reflections, the measured values may be merged, both confirming the symmetry and increasing the precision. Finally, for non-centrosymmetric crystals, especially those of chiral organic compounds, it is useful to measure at least some "Friedel" data, $\bar{h}\bar{k}\bar{l}$ (see section 10.2).

Control reflections. It is normal to choose 2–4 strong reflections the intensities of which are measured repeatedly at intervals during data collection. These check the consistency of the scattering power of the crystal. Some sensitive crystals decompose in the X-ray beam. The control reflections make it possible to correct for the fall-off in intensity. It is also usual to check the orientation of the crystal from time to time and, if necessary, redetermine the orientation matrix.

Intensity measurement. Once all of these preliminaries have been taken care of, automatic data collection can begin; one reflection after another is measured using the predetermined scan, and the data relating to it are written to a file. These data consist of the indices, the goniometer angles and the counts measured for the reflection and its background. For average crystals with normal X-ray generators, it is possible to measure 1000–2500 reflections per day. The measurement will then normally last between 1 and 14 days, depending on the size of the unit cell.

7.3 Area Detector Systems

A method long in use in protein crystallography involves the recording of many data at the same time by an *area detector*, which is the electronic equivalent of a photographic film. More recently, such devices have become more widely used for "small molecule" data collection. The first systems of this type used *multiwire proportional counters* as detectors. They consist of a mesh of wires in a chamber filled with xenon. Other early systems used picture tubes similar to those in TV cameras. There are two types of detectors currently in use:

- *Charge Coupled Device (CCD)-systems.* These use "CCD-chips" to record data, and are commonly found in digital cameras and camcorders. They are based on a layer of fluorescent material, such as gadolinium oxide sulfide sensitive to X-rays. For use with diffractometers, large CCD chips are used which have a diameter of 1–2 inches and a resolution of 1024×1024 or 4096×4096 pixels. Reflections can be recorded very quickly, almost in "real time". A problem is the level of background noise, which is reduced by cooling the chip with a Peltier element to between -40 and -60°C. This, however, makes them unsuitable for long exposure times required for weakly scattering material. Because of the small cross-section of the chip, the detector surface is magnified by a factor of about 1.5–3.6 using bundles of conical optical fibers. Since the resulting area, at most about 95×95 mm, is not large enough to encompass the required angular range for measuring all data in a single position, it must be mounted on a 3- or 4-circle diffractometer (Fig. 7.12).
- The *Image Plate* is also a new development. The plates currently used have diameters ranging from 180 to 350 mm (Fig. 7.13). Its surface consists of a layer of BaBrF, doped with Eu^{2+}. During exposure (typically 0.5–10 min.), incident X-ray quanta are converted to color centers (free electrons in interstitial lattice sites) resulting from the oxidation of Eu^{2+} to Eu^{3+}. This latent image is then read by a laser scanner, in much the manner a CD is read. The incident red laser beam causes the free electrons to reduce the Eu^{3+} back to Eu^{2+}, with the emission of photons in the blue-green region of the visible spectrum. The intensity of the emission by each pixel is then measured by a photocell with a photomultiplier. After full exposure to intense white light, to remove any possible remaining color centers, the plate is ready for reuse. The reading and regenerating phase requires 2–5 min. depending on the instrument. For this reason, despite the much larger diameter of the plate, image plate systems are usually somewhat slower than CCD-systems. Some apparatus has been designed using two or three image plates so that exposure of one plate can take place while another is being read back. Their great advantage is their very low background, as they are sensitive almost only to the scattered X-radiation. For

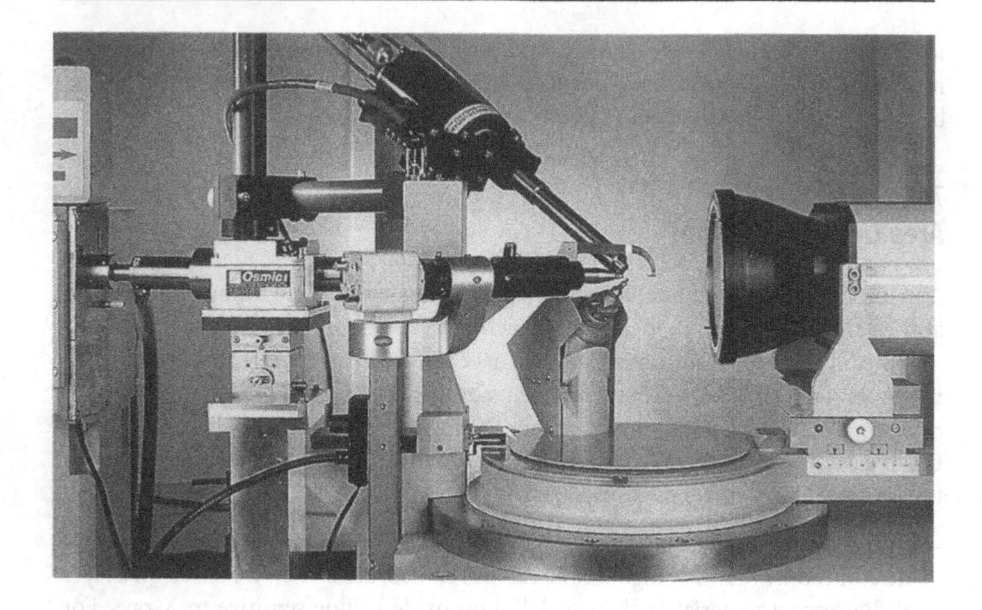

Fig. 7.12. Example of a 3-circle goniometer with a CCD-detector (Courtesy Bruker-Nonius AG).

this reason, weakly diffracting crystals, or crystals with weak super-structure reflections can be conveniently studied on image plate systems. Exposure times of up to

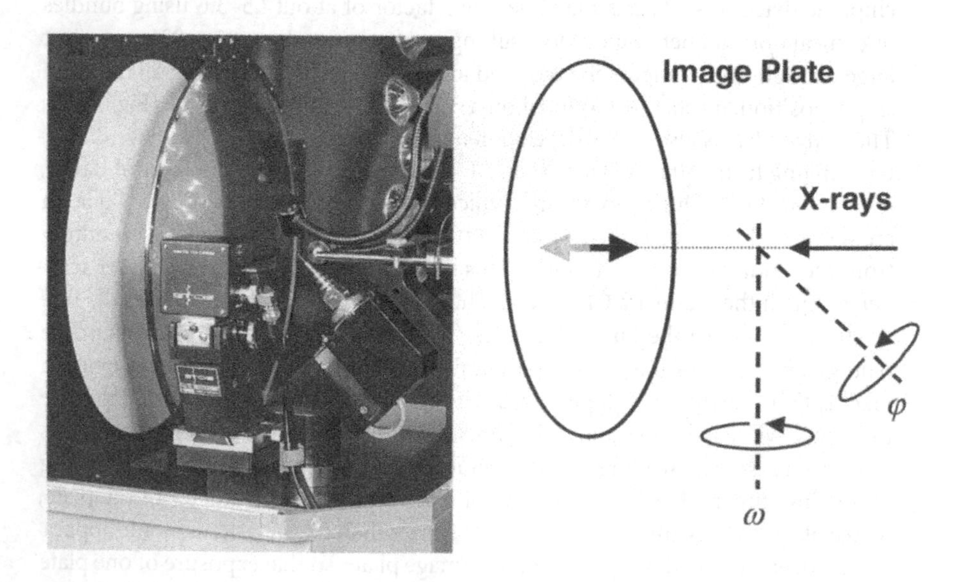

Fig. 7.13. Example of a two-circle diffractometer with an image plate (Courtesy Stoë & Cie.).

an hour are possible, as the half-life of the color-centers is about 10 hours. Because of the large diameter of the image plate, it is possible to measure an entire data set while rotating the crystal about a single axis. There is a small, funnel-shaped "dead area" about the rotation axis. Usually, though, 96–100 % of a data set can be measured, since symmetry-equivalent reflections will fill in the holes in the data. With two axes inclined to each other, 100 % "completeness" can usually be achieved. Image plate systems are simpler in construction than CCD-systems, and correspondingly less expensive. From the diameter of the plate and its distance from the crystal, the accessible angle range and the separation of reflections are mutually determined.

The sensitivities of image plates and CCD-systems are similar: they are about 50 times that of X-ray film, with a dynamic range of 10^5 and a resolution of about 50 × 50 μm. For "small-molecule" structures, only Mo-radiation is used with area detector systems, as it is difficult or even impossible to measure at scattering angles higher than 70°, which is necessary for copper radiation. On the other hand, Cu-radiation is better for macromolecular structures, where a smaller range of scattering angle is sufficient, but a better resolution of the data is essential. Since the required strategies for data collection are the same for the two systems, they are considered together here.

Data collection procedure. As with serial diffractometers, the crystal must first be centered precisely on the goniometer, and this is now assisted by a video camera. After that, a few orientation exposures are made which give information about the quality and diffracting power of the crystal and a tentative unit cell. This procedure is similar to the taking of rotation films (Fig. 7.14) except that the crystal is not orientated in any particular way. At the start, the crystal is rotated over a short angular range about the vertical axis. Typical rotation ranges would be from 0.5–2.0° for image plates or 0.3–1.0° for CCD-instruments. This brings a range of scattering vectors to the reflecting position, or, from the point of view of the reciprocal lattice, causes those reciprocal lattice points near the Ewald sphere to cut it — in Fig. 7.14, those in the gray area. Since such exposures give some 3-dimensional information, usually only a few are required to obtain the basic data needed to measure the whole diffraction pattern. (In the example in Chapter 15, only three exposures were needed: 0.1–1.2, 1.2–2.4 and 2.4–3.6°). A sample of an image plate exposure is in Fig. 7.15 and further one in Fig. 15.2.

Indexing. Each manufacturer has its own method for this, but all indexing programs are based on a peak search for scattering vectors over the first exposures, thus only 10–20 minutes from the start of measurement. On the basis of the differences between these vectors, reciprocal base vectors are sought, which can describe all found peaks as reciprocal lattice points. After Delauney reduction (section 2.2.2), a reduced cell is obtained for that originally found, and the corresponding orientation matrix is determined. From the metric symmetry of the cell, the probable crystal system can be selected.

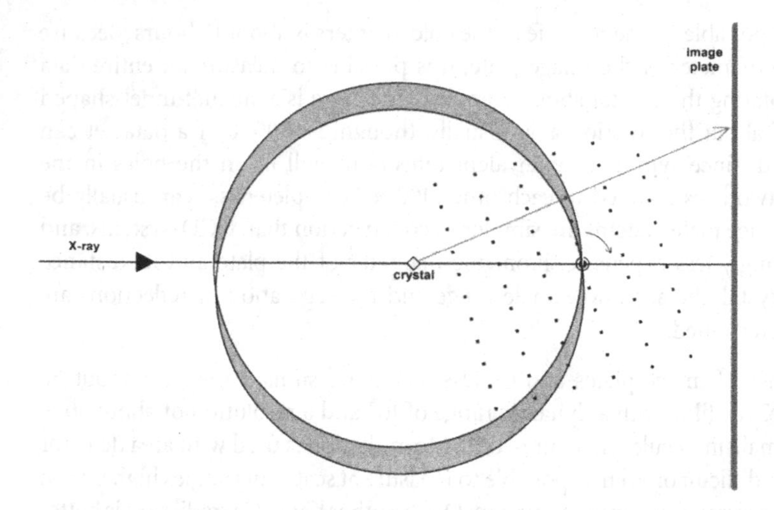

Fig. 7.14. Principle of the area-detector in terms of the Ewald construction. Rotation of, say, 1° about an axis normal to the plane of the paper brings the reciprocal lattice points in the gray area into coincidence with the Ewald sphere.

Measuring parameters. On the basis of these initially determined intensities and unit cell, suitable parameters for measurement can be chosen.

- *Exposure time.* This is chosen to make the strongest reflections close to the maximum measurable intensity of the pixels. For CCD-detectors, since the exposure time must be limited, multiple exposures are often needed.

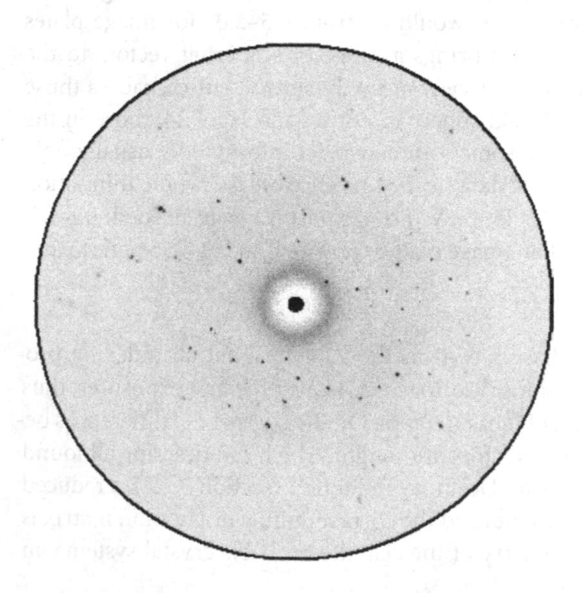

Fig. 7.15. Example of a 1° exposure of an area detector.

- *Rotation angle range.* The number of successive exposures is chosen to give at least one full set of independent data for the relevant Laue group. For image plate systems with a single rotation axis, this lies between about 150° for cubic and 250° for triclinic crystals. When there are several rotation axes, software specific to the apparatus can determine the optimum strategy. For example, two or more series of exposures over small ω-ranges can be made over different crystal orientations determined by the other goniometer axes. In any case, CCD-apparatus will require measurements to be made with more than one detector position if data at higher scattering angle are needed.
- *Detector to crystal distance.* The shorter the distance from the crystal to the detector, the greater is the accessible range of scattering angle, but the smaller the separation of the reflections ("spots") from one another. Thus, the setting of the detector will depend both on the size of the lattice constants and the breadth of the reflections. For image plates with fixed detector orientation, the shortest distance is chosen which gives the required scattering angle range while preventing spot overlap. With CCD-apparatus, somewhat longer measuring times are needed if both higher scattering angle data *and* greater spot separation are required.
- *Angle increment.* When the lattice constants are small, the separation of reciprocal lattice points is large, so relatively large ω-ranges may be used, say up to 2°, without danger of spot overlap. For larger cell dimensions, the rotation range must be correspondingly smaller. With image plates, as large a range as possible is chosen in order to minimize the number of exposures and save on reading time. For CCD-systems, steps of, say, 0.3–0.5° can be used, with the advantage of greater three-dimensional information, as more spots will occur on more than one exposure.

Representation of the diffraction pattern in reciprocal space. When enough exposures have been taken, a great advantage of area detector systems is that they display the entire diffraction properties of the crystal, not, as with serial diffractometers, merely where diffraction spots are expected. This can be used by selecting a particular cross-section of the reciprocal lattice, e.g. the $h0l$-layer. This will be built up by selecting pixels from all relevant exposures, and the resulting picture (e.g. Fig. 6.13) is the equivalent of a precession photograph (sect. 7.2.1). The spot-shape does appear somewhat distorted; this is particularly apparent when a large angle of rotation has been chosen for each step in data collection. Such sections of the reciprocal lattice are ideal for checking systematic absences, and can detect foreign reflections, twinning, satellites and diffuse reflection streaks, which, if present, will have to be considered in further treatment of the data.

Integration. If there are no anomalies in the data, the actual intensity measurement, that is the integration of the diffraction spots, can begin. The first step is to obtain a more accurate orientation matrix by selecting a range of spot positions through the data. Then, the software can determine the ϑ-dependent reflection profile for all the data. Given the new matrix and this profile information, the software can determine the exposures and the positions to search for components of each reflection *hkl*. Depending on the apparatus, an elliptical or a rectangular area is scanned, with an area determined by the profile-function. All the pixels inside the area are summed to the raw intensity, and those around the edge are used to estimate the background. They are then scaled and subtracted from the raw intensity to give the net intensity. Also, a standard uncertainty is determined for each intensity, and direction cosines are calculated for its peak position. Thereafter, all well-determined peak positions can be used for final refinement of the cell parameters. Using all the data in this way will tend to minimize any systematic errors arising from errors in centering the crystal or the apparatus.

7.4 Data Reduction

After data collection has been completed, the raw data, consisting of X-ray counts and measuring times for each reflection and its corresponding background, must be processed and corrected to produce F_o values, with which the calculated values (see Chapter 5) may be compared.

Net intensity. On serial diffractometers, the background region (in equation 7.1, B_L = left background and B_R = right background) is normally measured for about half the time taken to measure the peak region (P). If a given reflection has then been measured for a time t, the normalised net intensity (I) is given as:

$$I = [P - 2(B_L + B_R)]/t \tag{7.1}$$

This calculation is done along with the following in a "data reduction program". With area detector systems, this calculation will already have been done along with the integration.

7.4.1 *Lp* correction

The polarization factor. When electromagnetic radiation is reflected by a plane, the component of the beam with the polarization of its electric vector *parallel* to the plane is reflected without reduction of intensity. That portion *perpendicular* to the plane, however, is reduced in intensity by a factor of $\cos^2 2\vartheta$, i.e. by an amount dependent on the angle of reflection which becomes 0 for $\vartheta = 45°$. In total then, the *polarization factor p*, which is independent of the type of measurement, is

$$p = (1 + \cos^2 2\vartheta)/2 \tag{7.2}$$

When a graphite monochromator is used, the reflection by the monochromator affects the direct beam in this way before it reaches the crystal, and an additional factor K must be introduced to correct for this:

$$p = (1 + K\cos^2 2\vartheta)/(1 + K) \tag{7.3}$$

In fact, this correction is usually very small (for molybdenum radiation, the monochromator rarely makes a difference greater than 1 %) and it is often ignored.

The Lorentz factor. A further correction, this time dependent on the method of measurement, has to be made because different reflections remain in the diffracting position for different amounts of time. For a four-circle diffractometer, or for a zero-layer Weissenberg film, it is easily seen from the Ewald construction (Fig. 7.16), that for an ω-scan at constant angular speed, a shorter scattering vector will remain in the reflecting position for a shorter time than will a longer one, which moves more nearly tangential to the sphere. This angular dependent effect, called the *Lorentz factor L*, has the value:

$$L = 1/\sin 2\vartheta \tag{7.4}$$

Normally, these two effects are considered together as the so-called *Lp*-correction:

$$Lp = (1 + \cos^2 2\vartheta)/2\sin 2\vartheta \tag{7.5}$$

Thus, the observed structure factor F_o, on an arbitrary scale, may be calculated as:

$$F_o = \sqrt{I/Lp} \tag{7.6}$$

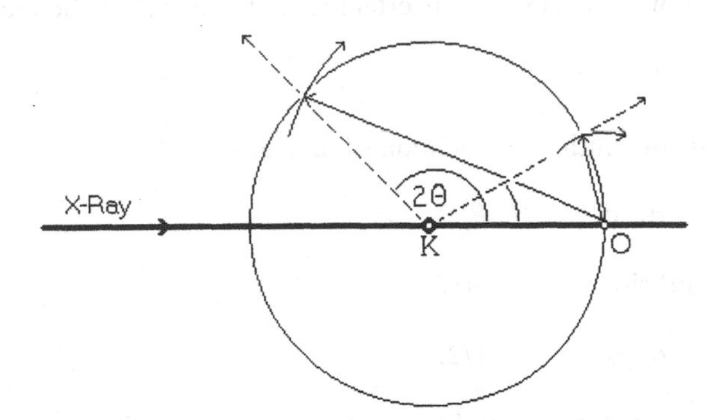

Fig. 7.16. Illustration of the Lorentz factor in terms of the Ewald construction.

With area detector systems, the fact that reflections in different parts of the reciprocal lattice are measured in different ways makes the *Lp*-correction somewhat more complicated.

7.4.2 Standard Uncertainty

At this stage the uncertainty in each measured intensity is estimated. Since the measurement of intensity using a counter is essentially the counting of X-ray quanta, the statistical error, or the standard uncertainty, is given simply as the square root of the number of counts

$$\sigma(Z) = \sqrt{Z} \tag{7.7}$$

Thus, the higher the number of counts, the higher its absolute standard uncertainty. The *relative* errors, however, become smaller. Note that the background intensities, which subtract from the intensity measurement *add* to its standard error, e.g. for a serial 4-circle diffractometer:

$$\sigma(I) = \frac{\sqrt{I_{\text{gross}} + 2(B_L + B_R) + (k \cdot I)^2}}{t} \tag{7.8}$$

A higher background thus raises standard uncertainties of the intensities significantly. In addition to this error, directly derived from counting statistics, an error for the variability of the instrument itself (kI, with $k \sim 0.01$–0.02) should be included. The normalization for time of measurement applied to the measurement of intensity must, of course, be applied also to its standard uncertainty. The corresponding value for the uncertainty of the Lp-corrected F_o^2 values is thus

$$\sigma(F_o^2) = \frac{\sigma(I)}{\sqrt{Lp}} \tag{7.9}$$

With area detector systems, the fact that reflections in different parts of the reciprocal lattice are measured under different geometrical conditions makes the Lp-correction somewhat more complicated. A rearrangement of equation 7.6 then allows the standard uncertainties of the F_o values themselves to be calculated:

$$I = LpF_o^2 \tag{7.10}$$

from which differentiation dI/dF_o gives:

$$dI/dF_o = 2LpF_o \tag{7.11}$$

and since $\sigma(I) = \sigma(F_o) \cdot |dI/dF_o|$

$$\sigma(F_o) = \sigma(I)/2LpF_o \tag{7.12}$$

This means that the relative error of the F_o values is half of those of the intensities or the F_o^2 values. This is significant when so-called "σ-cut-offs" are used. Sometimes, in the final cycles of least-squares refinement (see Chapter 9), weak reflections, say those with $F_o < 2\sigma(F_o)$ are not allowed to contribute. This "2σ-cutoff" in terms of F implies a "1σ-cutoff" in terms of I or F_o^2.

A problem, affecting the calculation both of the F_o values and of their standard uncertainties, occurs when the calculated value of a measured intensity is zero or negative. This arises, for weak or systematically absent reflections, where statistical variations make the background appear stronger than the reflection peak. One method of handling this is to replace all such measurements with an arbitrary positive value, say $\sigma/4$, in order to take the square root and thus obtain values of F_o and $\sigma(F_o)$ by equations 7.6 and 7.12. Since this clearly introduces systematic errors, it is better not to use F_o data but only F_o^2 data in refinement (see section 9.1).

Data reduction is normally carried out at the end of a data collection by the use of a simple program which then produces a file with the *hkl*-indices, the F_o^2-values, and their standard uncertainties, $\sigma(F_o^2)$. At the same time, the *direction cosines* can be calculated and saved. These give, for each reflection, the precise orientation of the crystal when it was measured.

7.4.3 Absorption Correction

As an X-ray beam passes through a crystal, it is weakened by elastic (Rayleigh) scattering, inelastic (Compton) scattering and by ionization. These effects increase roughly with the fourth power of the atomic number of a scattering atom and with the third power of the wavelength of the radiation. They can be summarized in terms of the *linear absorption coefficient,* μ by equation 7.13:

$$dI/I = \mu dx, \quad \text{or} \quad I = I_o e^{-\mu x} \tag{7.13}$$

The value of μ is the factor by which an element of the X-ray beam is weakened when it travels by a path x through the crystal. Values of μ are normally given in mm^{-1}, but care must be taken, as occasionally they are given in cm^{-1}.

The linear absorption coefficient for any compound may be calculated by adding up the contributions of the constituent atoms from the tabulated list of *Mass Absorption Coefficients* in International Tables C, Tab. 4.2.4.3 and multiplying by the calculated density of the crystals. This is done automatically by many program systems. Depending on the type of atoms present and the radiation used, the linear absorption coefficient normally has a value lying between 0.1 and 100 mm^{-1}. The type of correction to be made depends both on the magnitude of this value and on the size and shape of the crystal. Corrections become important if the crystal is very anisotropic — for example a very thin plate. In such cases, depending on the orientation of the crystal, there may be very great differences in the length of the path the beam must take through the crystal, and large errors in measured relative intensities may arise.

Numerical absorption correction. This is the best method, in which the lengths of all possible paths of the incident and scattered radiation through the crystal are calculated for each reflection. This is relatively simple for crystals with cylindrical or spherical shape. For the most exact measurements, e.g. for electron density measurements, it may be worthwhile to grind the crystal to a spherical shape. In normal cases, the crystal must be described by its bounding faces: From the orientation matrix, the *hkl* values for each face can be determined, and the perpendicular distance of these faces from a chosen point within the crystal can be measured (Fig. 7.17). These measurements are now usually made with the help of supplied software, and using the video camera supplied for crystal orientation on area detector systems. With it, it is possible to check that the calculated crystal faces actually correspond to the real ones, as in Fig. 7.18.

From the direction cosines (see above) for the measured data the actual orientation of the crystal at the time of measurement can be determined. The crystal is

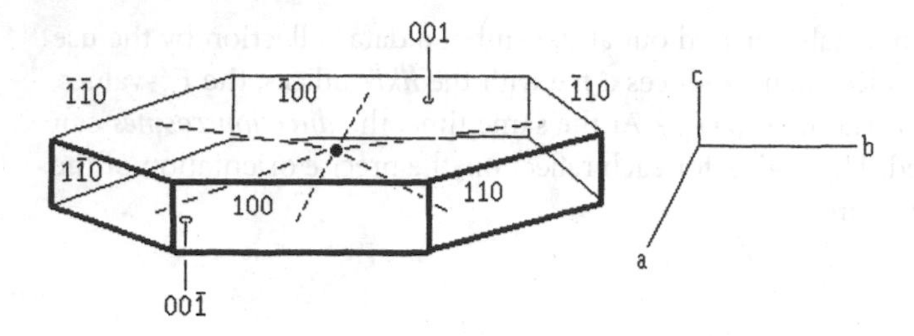

Fig. 7.17. Indexing and measurement of a crystal for an analytical absorption correction.

then divided into small elements of volume (at least 1000) and for each reflection, the path length of the direct and scattered beam through the crystal is calculated for each volume element. Integration over all of these (Gaussian integration) will give the reduction in intensity in the form of a *transmission factor* $A^* = 1/A$.

Semi-empirical absorption correction using ψ-scans. In many cases, the form of the crystal is very irregular, and the faces difficult or impossible to index. In any case, other absorption effects may be present, arising from adhesive or the glass rod used for mounting, which cannot be treated by the numerical method. In these cases, the semi-empirical method is particularly suitable. At the end of the data collection on a 4-circle diffractometer, a group of strong reflections are chosen, which are well distributed in reciprocal space and have χ-values near 90°. For these it is possible to measure complete *azimuthal* or *ψ-scans,* in which the reflection intensity is measured repeatedly (usually at 10° intervals) as the crystal is rotated about the normal to the actual lattice plane. This set of measurements gives an absorption profile for the crystal, since it changes in orientation during the rotation. With such measurements on a set of reflections (at least 6–10), a three dimensional absorption profile can be

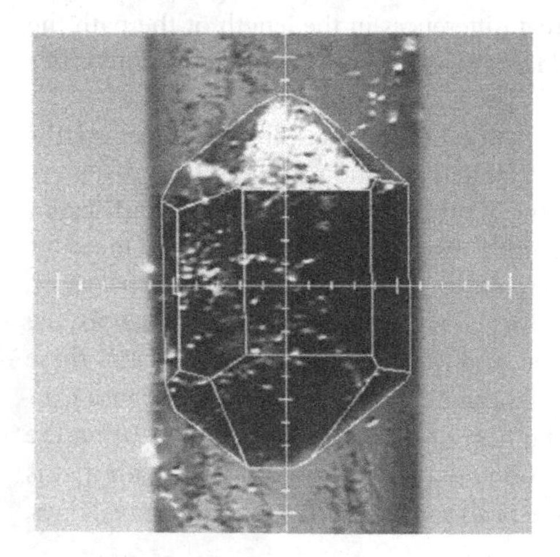

Fig. 7.18. Example of the description of a crystal by indexed faces.

calculated, and applied to the entire data set. A limitation to this method is that the requirement that only reflections with high χ-values can be used means that many possible crystal orientations are not adequately examined. It is better to use some partial ψ-scans obtained with lower χ-values. In some cases, extreme settings will mean that the beam is partly absorbed by the goniometer head. Such cases will be evident because of the unusually low background.

Semi-empirical absorption correction based on equivalent reflections. With measurements from area detector systems, a similar empirical correction can be made, if a sufficiently large range of reciprocal space is covered. It is particularly good in the case of higher Laue symmetries, when a considerable number of equivalent reflections, measured in different crystal positions, are present in the data set. Such a correction, based on several thousand reflections each with say 4–8 equivalents may well be better than a correction based on complete ψ-scans of only a few reflections.

The DIFABS method. A very successful, but also very controversial method [26] attempts to extract the absorption information from the differences between the F_o-data and the calculated F_c-values for the solved structure. This is done when a structure is solved and fully refined isotropically. The danger with this method is that differences between F_o- and F_c-values may also be caused by errors in the model being used for the structure. These are then "mopped up" in the absorption correction. It is wise to use this method with great care, and if possible to use it only when numerical or empirical methods are impossible, *e. g.* when the crystal has been destroyed.

The International Union of Crystallography (IUCr) requests authors for their journals to use a type of correction dependent on the value of μx, where x is the mean thickness of the crystal. For $0.1 < \mu x < 1$, a semi-empirical correction (based on ψ-scans or equivalent reflections) is acceptable; for $1 < \mu x < 3$, a numerical correction should be made if possible. For $\mu x > 3$, a numerical correction is always required.

7.5 Other Diffraction Methods

Crystals may also be investigated with various other radiations that have wavelengths similar to those of X-rays.

7.5.1 Neutron Scattering

There is a fundamental physical difference between the way X-rays and neutrons are scattered by matter. Neutrons are scattered by atomic *nuclei*, not by the electron shells. A result of this is that the scattering factors for neutrons are not proportional to Z as are those for X-rays, nor do they diminish greatly with scattering angle. Instead, they vary from one element to another, and in fact from one isotope of an element to another, and are independent of scattering angle. For example, the scattering factor of hydrogen is of average magnitude, so one important application of neutron scattering is the accurate location of hydrogen atoms, e.g. in hydrogen bonds. Another important application is the discrimination between atoms of similar atomic number, such as Co and Ni or Mn and Fe.

Neutrons are uncharged, but do have a magnetic moment, and this property has the unique advantage that scattering results not only for ordered arrays of atoms, but also from ordered arrays of magnetic moments. This means that neutron scattering is invaluable for the investigation of magnetic structures. For further details, see reference [28].

Neutron sources are available in many places. Leading laboratories for crystallographic use are the Institut Laue-Langevin (ILL) in Grenoble, France, and the Brookhaven National Laboratory in the United States. Neutron scattering generally involves relatively large samples (single crystals with volumes of 1–100 mm^3 or powder samples of a few grams) and long exposure times (typically 1–2 weeks per crystal). Polarized neutron scattering is also possible.

7.5.2 Electron Scattering

Electron beams interact strongly with matter, as they are scattered by both the nuclei and the electron shells of atoms. They are also strongly absorbed. Electron diffraction is thus very useful for measurements in the gas phase, and structure determination of small molecules with a few atoms is possible. Electron diffraction by solids is usually limited to high resolution transmission electron microscopy and recording the diffraction pattern of very thin flakes or films.

An interesting development for this [29, 30] has been the Fourier transform of the high resolution transmission image (the actual structure) to give phase information for the electron diffraction pattern (the reciprocal lattice) and thus a direct determination of structure, and even the refinement of it. There are many attendant problems, including the fact that in one crystal orientation only two dimensional data are usually accessible, and that the usefulness of the method is limited by damage to the crystal from the electron beam.

A big advantage of the method is that microscopically small ordered parts of the specimen are sufficient to give a scattering diagram. Thus, samples can be investigated for which no single crystals can be grown suitable for X-ray diffraction. There are, however, severe problems in measuring accurate intensities. In addition, problems arise in structure factor calculations because electron diffraction cannot be treated simply in terms of the kinematical scattering theory. It is difficult, therefore, to achieve structure refinements with R-values less than 20 %.

The use of synchrotron radiation as an alternative X-ray source is discussed in Ch. 3.1.

CHAPTER 8

Structure Solution

The direct results of the experimental measurements for a crystal structure are: the unit cell parameters, the space group (or at least a small selection of possibilities) and the intensity data. The real purpose of the work, however, remains — the location of the atoms in the unit cell.

8.1 Fourier Transforms

One way of looking at the diffraction phenomenon, from the point of view, so to speak, of the coherent X-ray beam, is that it is converted by the periodic array of electron density into the individual structure factors $F_o(hkl)$ by a process of *Fourier transformation.* A simple acoustical analogy is seen in the *Fourier analysis* of the sound of a violin string, or its decomposition into simple harmonic sine waves. If these individual waves, the *Fourier coefficients,* are known — both in amplitude and phase — they may be recombined by *Fourier synthesis* or *Fourier summation* in a synthesizer to give back the sound of the string. The situation is very similar in the case of X-ray diffraction. If all of the individual waves, that is the structure factors F_o with their phases, are known, a *Fourier synthesis* will give the electron density, and hence the crystal structure. The basic equations for this Fourier summation are:

$$\varrho_{XYZ} = \frac{1}{V} \sum_{hkl} F_{hkl} \cdot e^{-i2\pi(hX+kY+lZ)} \tag{8.1}$$

In this way the electron density ϱ_{XYZ} can be determined for every point XYZ in the unit cell (and, of course, only a single asymmetric unit needs to be calculated). In practice, it suffices to carry out the calculation for a grid of points with a spacing of 0.2–0.3 Å, and then by interpolation to locate the electron density maxima which correspond to the coordinates xyz of the atoms in the cell. This can be done manually by producing maps of the electron density on sections through the cell, and contouring them in the manner of geographical relief maps (see Fig. 8.1).

Since the experimental measurements have given only intensities, the phase information in the structure factors has been lost, and only their amplitudes are known. This is the underlying *phase problem* of crystal structure analysis, and is the theme of this chapter. The solution of a structure is, in fact, the solution of the phase problem.

The Structure Model. All the methods that will be described here depend sooner or later on the development of a *structural model,* that gives definite coordinates xyz

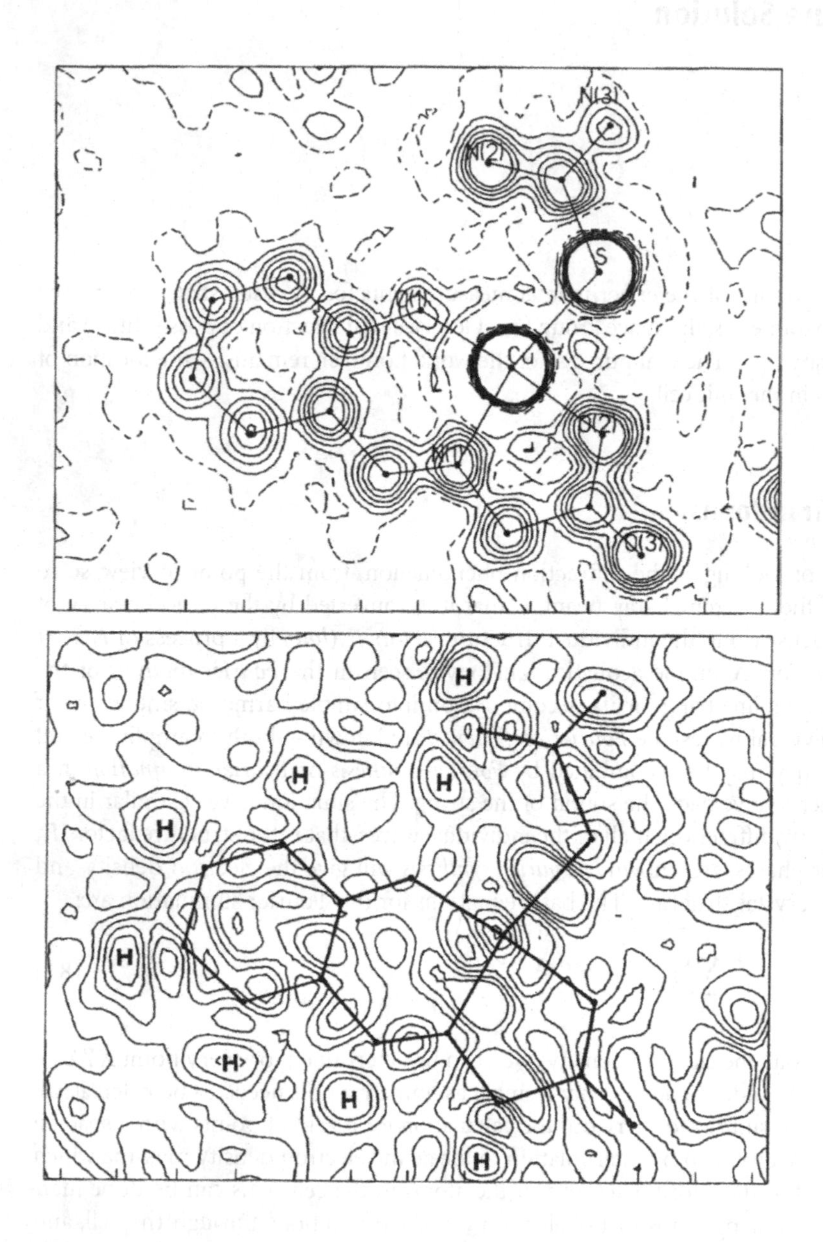

Fig. 8.1. a F_o-Fourier synthesis (above) and **b** difference Fourier synthesis: section through the molecular plane of the thiourea derivative of N-salicylideneglycinatocopper(II) ("CUHABS"—see Chapter 15). The contours are drawn at an interval of $1e\text{Å}^{-3}$ (**a**) and $0.1e\text{Å}^{-3}$ (**b**). The zero-contour is shown dashed. In **b**, not only the positions of the hydrogen atoms but also some of the bonding electrons in the rings can be seen.

relative to the origin of a definite space group for at least most of the atoms of the structure. If this model is more or less accurate, and contains sufficient structural information, it will permit theoretical structure factors, F_c to be calculated, using equation 5.9, (Chapter 5).

$$F_c = \sum_i f_i[\cos 2\pi(hx_i + ky_i + lz_i) + i \sin 2\pi(hx_i + ky_i + lz_i)] \tag{8.2}$$

These values, although they will contain definite errors, will also contain the required phase information. In particular, in centrosymmetric space groups, where the phase problem is essentially a sign problem, there is a high probability that a calculated sign is correct, although the calculated amplitude may be significantly in error. In general, it suffices that the model correctly describes about 30–50 % of the scattering matter in the asymmetric unit in order to have a useful set of calculated phases. The *calculated* phases can then be combined with the *observed* F_o values, and the resulting Fourier synthesis will reveal the entire structure, or at any rate a better model, on the basis of which the procedure can be repeated.

Difference Fourier Syntheses. In order to carry out the Fourier synthesis of equation 8.1, in principle all reflections *hkl* must be summed. In practice, only a limited data set can be measured, and this curtailing of the data will result in so-called series termination errors, which appear as ripples and false peaks in the electron density. This effect can be elegantly reduced by performing a summation in which the Fourier coefficients are the *observed* structure factors F_o (with the calculated phases) and then subtracting from each point an identical summation based on the *calculated* structure factors F_c for the same reflections. Since the series termination errors will be the same for both summations, they are effectively removed. A further advantage is that the peaks which remain represent the missing atoms of the structure, since the electron density for the model has effectively been subtracted from the "genuine" electron density. Such *difference* or ΔF *syntheses* are the usual method for the stepwise refinement of a partially known structure. They are particularly useful for location of hydrogen atoms in an organic structure which is otherwise complete. Fig. 8.1 shows an F_o synthesis (above) and a ΔF synthesis (below) on the basis of a model consisting of the non-hydrogen atoms in the asymmetric unit.

It is now necessary to examine the methods by which such a model can be deduced in the first place.

8.2 Patterson Methods

The method introduced by *Patterson* for the determination of a structural model makes use of a Fourier synthesis like that in equation 8.1. In the *Patterson function* P_{uvw}, however, the coefficients are the directly measured F_o^2 values. In order to distinguish it from a "normal" F_o Fourier synthesis, the coordinates in *Patterson space* are given the symbols *u*, *v*, *w*. Although they refer to the same axes and unit cell, these coordinates are *not* directly related to atomic coordinates x, y, z.

$$P_{uvw} = \frac{1}{V} \sum_{hkl} F_{hkl}^2 \cdot \cos[2\pi(hu + kv + lw)] \tag{8.3}$$

Since the F_o^2 values contain no phase information, the Patterson function contains only that information which is implicit in the intensities themselves, namely the *interatomic vectors*. As was explained in section 5.3, the result of summing the waves corresponding to the amplitudes alone – and thus also to that for the intensities alone – is that only the relative interatomic separations are retained. These depend only on the components of the interatomic vectors in the direction of the scattering vector $\boldsymbol{d}^*(hkl)$. It is only when phase information is supplied that the vectors are related to a particular unit cell origin. In other words, a Fourier synthesis using *only* intensities will contain *only* interatomic vectors, each originating from a single point, the origin of the Patterson function.

When the Patterson function has been calculated for the whole unit cell, it will contain maxima, corresponding to the ends of the interatomic vectors. Such a result is shown diagramatically in Fig. 8.2 for a two atom structure with a symmetry center. It can be seen that each interatomic vector operates in both directions, and that the presence of the inversion center makes vector 1 and 2 have double weight.

Intensity of a vector. The relative intensity of a Patterson maximum is easily estimated as the product of the two numbers of electrons, i. e. the two atomic numbers.

$$I_P = Z_1 \cdot Z_2 \tag{8.4}$$

The origin of the cell will always correspond to the highest peak, since every atom is at a distance 0 from itself, and so the peak will have a height proportional to the sum of the squares of the atomic numbers of all the atoms in the cell. When Patterson peaks are listed by a program, the origin peak is often arbitrarily scaled to a value of 999; all of the peaks can then be normalized by the scale factor:

$$k = \sum \frac{Z_i^2}{999} \tag{8.5}$$

It can easily be seen that the Patterson function for a large organic molecule, with many atoms of roughly equal atomic number will be very complex and difficult to interpret. When there are a few atoms that are much heavier than the rest, as is

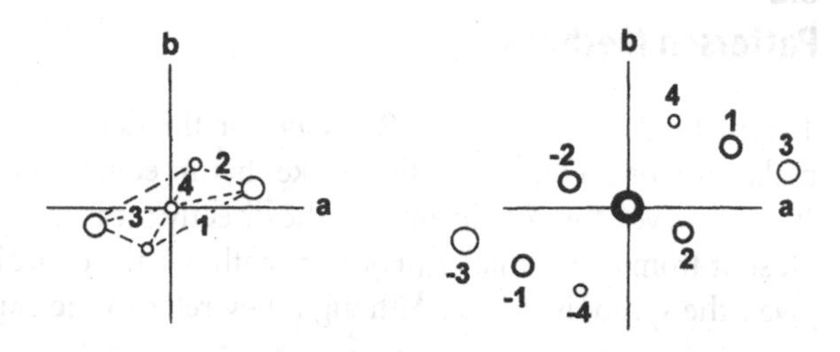

Fig. 8.2. The derivation of maxima in a Patterson synthesis. Double weight maxima are printed in bold.

usual for typical organometallic compounds, the vectors between these atoms will be prominent, while those corresponding to separations of two light atoms will disappear into the background. Table 8.1 gives an example for the peaks of $(C_5H_5)_3Sb$.

8.2.1 Symmetry in Patterson Space

The symmetry elements relating atoms to one another in space must also be apparent in the Patterson function. For example, if there is a 2-fold axis through the origin parallel to $\boldsymbol{b}$, every atom at a point x, y, z is related to a second at $\bar{x}, y, \bar{z}$. The vectors separating these atoms will be at $2x, 0, 2z$ and $-2x, 0, -2z$. Peaks like these, between atoms related by symmetry, are called *Harker peaks.* In the case of the 2-fold axis as above, a result will be that there is a particularly prominent set of vectors in the $u0w$ plane, which is called a *"Harker plane"*. If, instead, there is a mirror plane normal to $\boldsymbol{b}$, all of the atoms related by it will give peaks on the $0v0$ line, the *"Harker line"*. In fact, a Patterson map is often a way of locating those symmetry elements which are not manifest in the Laue symmetry. One problem is that a heavy atom may actually lie on a symmetry element, so that the vectors characteristic of that element are not found. In general, when a Harker line is being sought, it is better to look at a contour diagram rather than a peak list, since the presence of many superimposed vectors may result in there being few maxima.

Since all interatomic vectors operate equally in both directions, a Patterson map is always centrosymmetric. The translation vectors in a centered lattice are also present in its Patterson function, and so the Bravais lattice of the Patterson is the same as that of the structure. Screw axes and glide planes, however, are converted to the corresponding rotation axes and mirror planes in Patterson space. The translation component is however revealed by the position of the resulting Harker plane or line.

Example: The c-glide plane normal to $\boldsymbol{b}$ *in space group* $P2_1/c$ *transforms an atom at* x, y, z *into one at* $x, \frac{1}{2} - y, \frac{1}{2} + z$. *The vectors relating these two atoms are at* $0, \frac{1}{2} - 2y, \frac{1}{2}$ *and* $0, \frac{1}{2} + 2y, \frac{1}{2}$. *The two peaks are thus related in Patterson space by a mirror plane at* $x, \frac{1}{2}, z$ *and the c-glide component is clearly indicated by the fact that the peaks both have* $w = \frac{1}{2}$.

8.2.2 Structure Solution Using Harker Peaks

Symmetry can also provide the key to the "solution" of a Patterson synthesis, that is the determination of the coordinates x, y, z for an atom in the structure model by interpreting the Patterson maxima. This is illustrated in Table 8.1 by the compound $(C_5H_5)_3Sb$, which crystallizes in the space group $P2_1/c$ with $Z = 4$ formula units per unit cell.

The general position in space group $P2_1/c$ is 4-fold, and it is highly probable that the Sb-atom occupies a general position rather than two independent inversion centers, which would in any case be impossible for a molecule of the expected structure.

Table 8.1. Determination of the positional parameters x, y, z for Sb in $(C_5H_5)_3$ Sb from the Harker peaks of a Patterson synthesis.

a) Patterson normalization based on the height of the origin peak (there are four $(C_5H_5)_3$Sb-units per cell.)

n	Atom	Z	Z^2	nZ^2	
4	Sb	51	2601	10404	
60	C	6	36	2160	$f = 999/12564 = 0.0795$
				12564	

Calculated peak heights ($f \cdot Z_1Z_2$): Sb–Sb 207 Sb–C 24 C–C 3

b) Harker-peaks in space group $P2_1/c$.

	x, y, z	$\bar{x}, \bar{y}, \bar{z}$	$\bar{x}, \frac{1}{2}+y, \frac{1}{2}-z$	$x, \frac{1}{2}-y, \frac{1}{2}+z$
x, y, z	–	$-2x, -2y, -2z$	$-2x, \frac{1}{2}, \frac{1}{2}-2z$	$0, \frac{1}{2}-2y, \frac{1}{2}$
$\bar{x}, \bar{y}, \bar{z}$	$2x, 2y, 2z$	–	$0, \frac{1}{2}+2y, \frac{1}{2}$	$2x, \frac{1}{2}, \frac{1}{2}+2z$
$\bar{x}, \frac{1}{2}+y, \frac{1}{2}-z$	$2x, \frac{1}{2}, \frac{1}{2}+2z$	$0, \frac{1}{2}-2y, \frac{1}{2}$	–	$2x, -2y, 2z$
$x, \frac{1}{2}-y, \frac{1}{2}+z$	$0, \frac{1}{2}+2y, \frac{1}{2}$	$-2x, \frac{1}{2}, \frac{1}{2}-2z$	$-2x, 2y, -2z$	–

c) The strongest peaks in the Patterson synthesis.

№	height	u	v	w	assignment	
1	999	0	0	0	zero peak	
2	460	0	0.396	0.5	Harker peak $0, \frac{1}{2}-2y, \frac{1}{2}$	2 × Sb–Sb
3	452	0.420	0.5	0.705	Harker peak $2x, \frac{1}{2}, \frac{1}{2}+2z$	2 × Sb–Sb
4	216	0.421	0.104	0.206	Harker peak $2x, 2y, 2z$	1 × Sb–Sb

d) Calculation of the Sb-position from correspondence of (b) and (c).

peak 2:	$\frac{1}{2}-2y$	=	0.396	$\Longrightarrow$	$y = 0.052$
peak 3:	$2x$	=	0.420	$\Longrightarrow$	$x = 0.210$
	$\frac{1}{2}+2z$	=	0.705	$\Longrightarrow$	$z = 0.103$

It is then easy to calculate algebraically the Harker peaks – the interatomic vectors relating the symmetry equivalent atoms. It will be seen that some of them have double weight and lie on *Harker planes* (one constant parameter) or *Harker lines* (two constant parameters) in Patterson space. Examination of the table of peaks shows that the three strongest peaks – after the origin – can be interpreted such that the u, v, w values for all three are consistent with a single value of x, y, z for Sb. This model, although it knows nothing about the positions of the 15 C and 15 H atoms does contain 33 % of the electrons of the structure. Since these are so strongly localized in the heavy atom, their contribution to the F_c-values is great enough to allow nearly all of the signs of the reflections to be calculated, and a subsequent difference

Fourier synthesis reveals all of the C-atom positions as clear peaks. This technique, whereby the localisation of a large part of the electron density in one or a few heavy atoms makes the Patterson function easy to interpret, is known as the *heavy atom method.*

*A method which has found great application in macromolecular (mostly protein) crystallography, the "isomorphous replacement method" is an extension of the same principle. For it, crystals are grown — so called "derivatives" — in which various heavy atoms (Br, I, metals) are introduced into various sites without significantly altering the unit cell dimensions or the positions of the rest of the atoms. A complete data set is measured for each crystal, including one with no heavy atoms added — the "native" data. The heavy atoms are located by Patterson maps, often making use of a crude but effective subtraction of the native data from that of the derivative, and thus some phase information is obtained. From two or more such derivatives it is often possible to obtain enough phases to solve the native data set (**M**ultiple **I**somorphous **R**eplacement, **MIR** method). At present, increasing use is made instead of the differences in anomalous dispersion possible by measuring the same crystal containing a heavy atom like Se, Br at different wavelengths. The choice of wavelengths above and below absorption edges, possible now because of the availability of synchrotron radiation (see section 3.1), provides maximum differences in anomalous dispersion effects (**M**ulti-Wavelength **A**nomalous **D**ispersion, **MAD** method). Much use is also made in protein crystallography of the method discussed in the next section.*

8.2.3 Patterson shift methods

The solution of a structure entirely by the interpretation of Harker peaks is usually only practicable when there are one or two heavy atoms in the asymmetric unit which differ significantly in scattering power from the rest of the atoms. In the case where the structure contains groups of similar atoms of known stereochemistry, another technique for interpreting the Patterson is possible, which will only be briefly mentioned here. The Patterson function for such a fragment can be calculated by considering the vectors resulting from placing each atom in turn at the origin, as illustrated in Fig. 8.3. Such fragments such as rigid ring systems occur in many structures — chemists do, after all, often know what to expect! The problem is then to find the orientation and the position of the fragment in the unit cell by comparing this calculated vector map with the actual Patterson function. There are several programs which automate this "image-seeking" procedure, usually by first performing a "rotation search" for the orientation and then a "translation search" for the position [e.g. 61, 65, 67]. The location of a suitably large fragment makes possible the completion of the structure by difference Fourier methods. In the "superposition method," used e.g. in SHELXS-97, a single Harker vector is found (e.g. $2x, 2y, 2z$ in the example of Tab. 8.1). A duplicate of the original Patterson map, shifted by this vector, is then superimposed on it. If the "minimum function" at each point of the map is calculated, the structure becomes

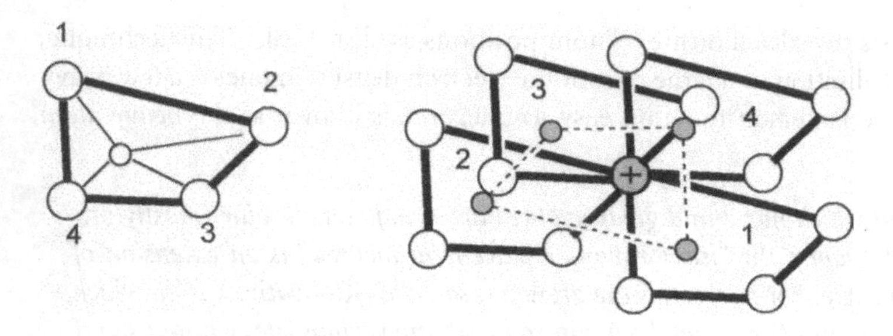

Fig. 8.3. Illustration of the Patterson image-seeking method.

visible (if the right Harker vector was chosen) in this map and its position relative to the origin can be calculated.

The Patterson method is difficult to apply when the structure consists of many atoms of roughly equal scattering power and no known stereochemistry. In such cases, an alternative approach is more suitable.

8.3 Direct Methods

The methods to be described here are called *direct* because they make use of relationships between the intensities of the various reflections which lead directly to a solution to the phase problem.

8.3.1 Harker-Kasper Inequalities

Direct methods may be said to date from 1948, when Harker and Kasper showed that as a result of the symmetry elements, useful relationships arise between the amplitudes of specific pairs of reflections. Instead of the measured structure factors, they employed *unitary structure factors* U, which have been normalized by use of $F(000)$, the total number of electrons in the unit cell:

$$U = F/F(000) \tag{8.6}$$

They give an indication what fraction of the electrons are scattering in phase for that reflection. An important example may be derived for the center of symmetry in space group $P\bar{1}$: the resulting inequality is:

$$U^2_{hkl} \leqslant \frac{1}{2} + \frac{1}{2} U_{2h\,2k\,2l} \tag{8.7}$$

This means that whenever U^2_{hkl} is very large, i. e. greater than 2, the second order of this reflection $U_{2h\,2k\,2l}$ must have a positive sign, and this must be increasingly probable as the values of U for the two reflections increase. This and similar inequalities are not sufficient on their own to define enough phases to solve the structure. Later, however, more generally useful relationships were found, also relating the relatively strong reflections of a structure.

8.3.2 Normalized Structure Factors

Measured structure factors have magnitudes that fall off strongly with ϑ, the scattering angle, because of the ϑ-dependence of the atomic scattering factors. This makes it difficult to compare them with one another, but the effect can be corrected by normalizing F^2 by means of the "expectation value" for its scattering angle, essentially the mean value for that range of ϑ: $\langle F^2 \rangle$:

$$E_o^2 = \frac{F_o^2}{\langle F^2 \rangle} \tag{8.8}$$

The expectation value, on an absolute scale, assuming no atomic vibrations, can be estimated by *Wilson statistics* as the sum of the squares of the atom scattering factors for the given scattering angle. Since the ϑ-dependence of F^2 is also dependent on the atomic vibrations, an "overall" displacement factor as well as a scale factor for converting the intensities to an absolute scale can be estimated by comparing and correcting this for the mean displacement factor as in equation 5.4:

$$\langle F^2 \rangle = k\varepsilon \sum f_i^2 \exp(-2B \sin^2 \vartheta / \lambda^2) \tag{8.9}$$

(For some classes of reflections, a "symmetry enhancement factor," ε, is required. This is a small integer – for more information, see Int. Tab. B Table 2.1.3.3). This can be written in a convenient linear form:

$$\ln \left(\frac{\langle F^2 \rangle}{\varepsilon \sum f_i^2} \right) = -\ln k + 2B \sin^2 \vartheta / \lambda^2 \tag{8.10}$$

whereby it will be seen that the calculation of normalized structure factors has a bonus in giving approximate values of both the scale factor and the overall displacement factor for the data.

Since the expectation values are measured by averaging measured data in the relevant ϑ-range, it is important that *all* possible data are measured and included in the data set, including the very weak reflections.

Statistics of E-values. An E-value greater than 1 indicates one which is greater than the expectation value. The strongest few percent of data have $|E| > 2$. It can be shown that for centrosymmetric structures, the statistical probability for very large or very small E-magnitudes is greater than for non-centrosymmetric structures. This is because in a non-centrosymmetric structure, the vectorial addition of the contributing atom scattering factors tends to give resulting magnitudes grouped around a medium value, zero and very large intensities are improbable. In centrosymmetric structures, on the other hand, the imaginary parts of the scattering factor contributions are cancelled by the atom pairs related by the center. The vectorial addition of the real parts alone may lead with a much higher probability to $F = 0$ or large values. The difference is easily seen in the resulting mean value of $E^2 - 1$ for all data. For non centrosymmetric structures, it has a value near 0.74, while for centrosymmetric it is about 0.97. This is very useful in space group determination: when the Laue symmetry and the systematic absences leave a choice between two or more space groups, the most common situation is that a choice must be made between a centrosymmetric and a non-centrosymmetric group. Typical examples of such "pairs" are $Pnma$ and $Pn2_1a$ ($= Pna2_1$), $P2_1$ and $P2_1/m$, $C2/c$ and Cc, and $C2/m$, Cm and $C2$. In such cases, the calculation of the value of $\langle E^2 - 1\rangle$ for the structure may indicate which space group is more likely. If the value does not lie near one of these values, a very low value may be a sign of twinning (see section 11.2), while an abnormally high value may indicate "hypersymmetry," the situation where the asymmetric unit of a centrosymmetric structure has itself a non-crystallographic center of symmetry. These approximate centers will exaggerate the centric distribution of intensities and hence raise the value of $\langle E^2 - 1\rangle$ even higher.

8.3.3 The Sayre Equation

A very important step in the development of direct methods was the formulation by *Sayre* of a relationship based only on these two assumptions: that the electron density in a structure can never have a negative value, and that it is concentrated in well defined maxima:

$$F_{hkl} = k \sum_{h'k'l'} F_{h'k'l'} \cdot F_{h-h',k-k',l-l'} \tag{8.11}$$

This equation states that the structure factor for any reflection hkl can be calculated as the sum of the products of the structure factors of all pairs of reflections whose indices sum to it, e. g.

$$E_{321} = E_{100} \cdot E_{221} + E_{110} \cdot E_{211} + E_{111} \cdot E_{210} + \cdots \tag{8.12}$$

At first glance, this seems of little use, since the calculation of a single reflection requires the knowledge of many others, complete with phase information. It must be remembered, however, that the contribution of any pairs of which one or both are weak will make little contribution and can, at least to begin with, be ignored. As the values increase, so does the importance of their contribution. A result of this is that if both reflections in the pair are very strong *and* the reflection being calculated is strong as well, then there is a high probability that this relationship alone will give a good indication of the phase of E_{hkl}.

8.3.4 The Triplet Relationship

It is above all due to Karle and Hauptman that this principle was developed into a practical method which is used today more than any other method for structure solution. It was mainly for this, that they received the Nobel Prize for Chemistry in 1985. For centrosymmetric structures, in which the phase problem is reduced to a "sign problem," equation 8.11 implies that, for a so-called Σ_2 triplet of strong reflections contributing to the Sayre equation:

$$S_H \approx S_{H'} \cdot S_{H-H'} \tag{8.13}$$

(For convenience, here and in the following paragraphs, hkl is represented by H and $h'k'l'$ by H'.) For example, the relationship $S_{321} = S_{210} \cdot S_{111}$ implies that if the reflections 210 and 111 both have the same sign (positive or negative) it is probable that the sign of 321 is positive. On the other hand, if 210 and 111 have opposite signs, it is likely that 321 will have a negative sign. The underlying principle is easy to visualize. If *all* the atoms lie on a set of lattice planes, then they will all scatter in phase, and that reflection will be outstandingly strong (cf. section 5.3). If the origin is taken to lie on the planes, they will have a phase angle of 0°, or a sign of + (Fig. 8.4 left). Put the other way around, a strong reflection with a phase angle of 0° will, in a Fourier synthesis, contribute substantial electron density at the crests of its lattice planes. On the other hand, if the atoms lie on planes parallel to and midway between pairs of these lattice planes, (Fig. 8.4 right), the reflection will also be very strong, but will have a phase of 180°, or a sign of −. And, of course, this implies that such reflections will contribute in a Fourier synthesis to the electron density on the planes at $d/2$, $3d/2$, etc.

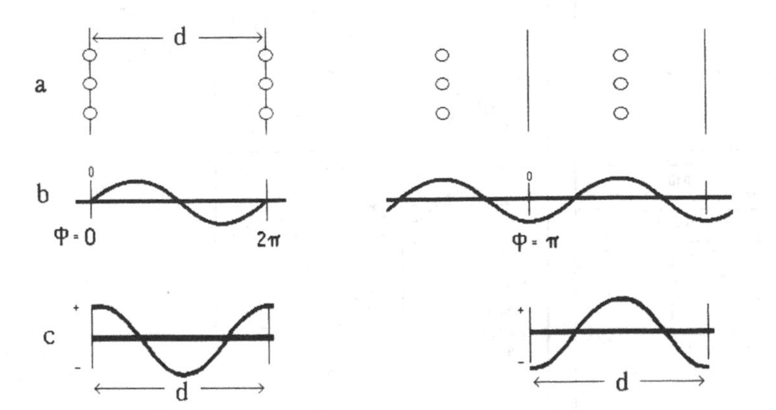

Fig. 8.4. a) left: atoms *on* the planes **right:** *between* the planes of a set of lattice planes. **b)** Waves diffracted by these planes: **left:** with phase of 0(+). **right:** with phase $\pi(-)$. **c)** Contribution of the reflection from this set of lattice planes to the Fourier synthesis, along the normal to these planes.

When three strong reflections have a triplet relationship to one another: H, H' and $H - H'$ (it e.g. 110, 100 and 010), then the above argument applies to all of them *at the same time,* and electron density must be concentrated at intervals of d for all three sets of lattice planes. If the phases of two of the reflections are known, then the phase of the third reflection is determined by them, as is shown in Fig. 8.5.

Each Σ_2-relationship in fact gives the probability that the sign of a reflection H is determined by those of H' and $H - H'$, the probability depending on how strong the three reflections are. Cochran and Woolfson [32] demonstrated that, for a structure with N atoms of equal weight in the unit cell, the probability p, that a phase is correctly determined is given by the relationship:

$$p = \frac{1}{2} + \frac{1}{2}\tanh\left[\frac{1}{\sqrt{N}} E_H E_{H'} E_{H-H'}\right] \tag{8.14}$$

It is thus clear that direct methods work less well in principle for more complex structures. The limits for the method currently lie in the region of 200–300 non-hydrogen atoms in the asymmetric unit.

The Σ_1-relationship. This is a special case of the Σ_2-relationship, in which the reflections H' and $H - H'$ are identical, e. g. such a relationship as:

$$S_{222} \approx S_{111} \cdot S_{111} \tag{8.15}$$

This implies that the sign of a reflection $2h\,2k\,2l$ is *positive,* if both the reflection hkl and its second order $2h\,2k\,2l$ are strong, independent of the sign of hkl. This is, of

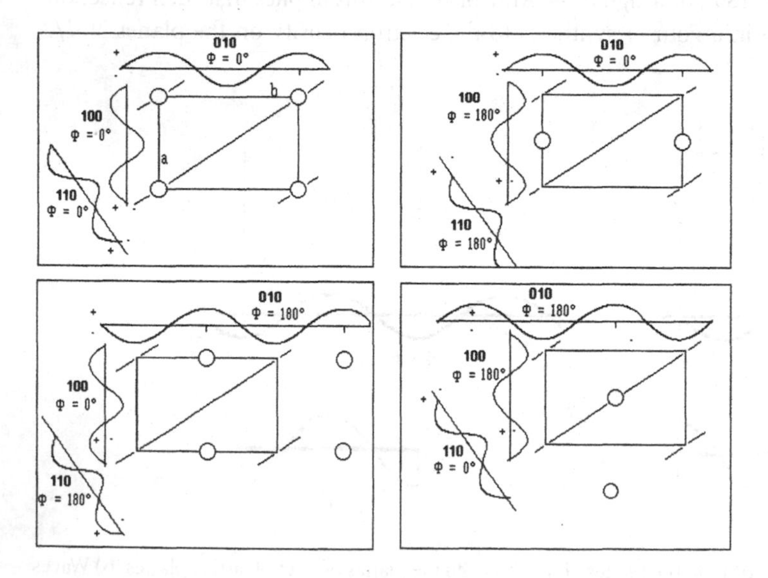

Fig. 8.5. Correlation between phases and atom positions for the reflection triple 010, 100 and 110. Contribution of these reflections to the Fourier synthesis along the direction normal to the planes.

course, simply an expansion of the Harker-Kasper inequality (eq. 8.7). Any such pair of reflections — which are relatively uncommon — thus gives an indication, with an associated probability, of the sign of a reflection.

Noncentrosymmetric structures . When there is no center of symmetry in a structure, then instead of a sign, a phase angle Φ must be determined. In the general form of the Sayre equation (Eq. 8.11) the structure factors are complex quantities, and an analogous to the sign relationship, a phase relationship may be derived:

$$\Phi_H \approx \Phi_{H'} + \Phi_{H-H'} \tag{8.16}$$

The objective is to derive a phase Φ_H, normally from a large number of Σ_2-relationships. Karle and Hauptman derived for this purpose the so-called *tangent formula,*

$$\tan \Phi_H = \frac{\sum_{H'} \varkappa \cdot \sin(\Phi_{H'} + \Phi_{H-H'})}{\sum_{H'} \varkappa \cdot \cos(\Phi_{H'} + \Phi_{H-H'})} \tag{8.17}$$

in which the summation is carried out over all relevant triplets. The size of $\varkappa = N^{-1/2}|E_H E_{H'} E_{H-H'}|$ is related to the probability of each relationship analogously to equation 8.14.

In addition to the triplet-relationship, more complex relationships have been derived. A particularly important one is the *quartet* relationship:

$$\Phi_4 \approx \Phi_{H1} + \Phi_{H2} + \Phi_{H3} + \Phi_{H1+H2+H3} \tag{8.18}$$

If the product of all these E-values $E_{H1}E_{H2}E_{H3}E_{H1+H2+H3}$ is large, then the intensities of the "cross terms" E_{H1+H2}, E_{H2+H3} and E_{H1+H3} determine the probable value of Φ_4. If they are strong, then Φ_4 has a value near 0°, and the relationship is called a *positive quartet* and the relationship is simply a combination of triplets. More significant is the result when the cross terms are weak. In this case, Φ_4 has a value near to 180°, and the relationship is a *negative quartet.* Such relationships are useful not only in the determination of new phases but also in the evaluation of the probability of various alternative solutions to the phase problem (see below).

8.3.5 Origin Fixation

In a centrosymmetric structure, there are other centers of symmetry than the one at the origin (cf. Ch. 6, Fig. 6.5). There are also inversion centres at the center of each edge ($\frac{1}{2}$, 0, 0; 0, $\frac{1}{2}$, 0; and 0, 0, $\frac{1}{2}$) at the center of each face (0, $\frac{1}{2}$, $\frac{1}{2}$; $\frac{1}{2}$, 0, $\frac{1}{2}$; and $\frac{1}{2}$, $\frac{1}{2}$, 0) and at the center of the cell ($\frac{1}{2}$, $\frac{1}{2}$, $\frac{1}{2}$). In the triclinic, monoclinic and orthorhombic systems, it is possible to describe the structure equally well in eight different ways by choosing any one of these points as the origin of the cell. For example, it is permissible to add 0.5 to all the x-parameters of the atoms. Such a change will have no effect on the amplitudes of the scattered waves, but will affect their phases, which will also be shifted (see Table 8.2). This may be demonstrated by considering the reflection 100. If an atom determining the phase is at the origin, the phase will be 0° and the sign positive. Shifting the origin to $\frac{1}{2}$, 0, 0 makes the phase 180° and the sign negative. By the selection of a positive sign for this reflection, the origin has been fixed in the first position. In the same way, suitable "independent" reflections may have their signs fixed in order to fix the origin along the other two axial directions.

Table 8.2. Phase shifts resulting from origin shifts.

origin \ class	*ggg*	*ugg*	*gug*	*ggu*	*uug*	*ugu*	*guu*	*uuu*
0, 0, 0	+	+	+	+	+	+	+	+
$\frac{1}{2}$, 0, 0	+	−	+	+	−	−	+	−
0, $\frac{1}{2}$, 0	+	+	−	+	−	+	−	−
0, 0, $\frac{1}{2}$	+	+	+	−	+	−	−	−
$\frac{1}{2}$, $\frac{1}{2}$, 0	+	−	−	+	+	−	−	+
$\frac{1}{2}$, 0, $\frac{1}{2}$	+	−	+	−	−	+	−	+
0, $\frac{1}{2}$, $\frac{1}{2}$	+	+	−	−	−	−	+	+
$\frac{1}{2}$, $\frac{1}{2}$, $\frac{1}{2}$	+	−	−	−	+	+	+	−

The three reflections required to fix the origin should be strong ones which contribute to many Σ_2-relationships. The Σ_2-relationship may be written in the alternative form:

$$\Phi_3 = \Phi_{H1} + \Phi_{H2} + \Phi_{H1+H2}$$

where Φ_3 has a probable value of 0. Such values are *entirely* independent of the choice of origin, and they are thus *structure invariants,* although the values of the phases from which they are derived will usually shift with the choice of origin. Reflections whose indices are all even (class *ggg* in Table 8.2) are unaffected, since any origin shift of $\frac{1}{2}$ will cause a phase shift of $2\pi = 360°$. (This may be demonstrated by substituting $hkl = 222$ in equation 8.14.) Such reflections cannot be used to help fix the origin and are called *structure seminvariants,* as their phase is independent of the choice of origin, *so long as it is chosen to lie on a center of symmetry.*

8.3.6 Strategies of Phase Determination

All of the important methods for structure determination by direct methods begin from a starting set of reflections with *known phases.* Then a list is made of a few hundred other reflections with large E-values, some of whose phases can be estimated with high probability from the known phases in the starting set. The procedures from this point on vary from one program system to the next (see below). As soon as enough E-values have phases determined, a Fourier synthesis based on them, like one based on the F-values, should reveal the positions of all or nearly all of the atoms in the structure so that the structure can then be completed by difference Fourier methods.

Starting sets. The starting set should fix the origin completely, and may contain some reflections whose phases are satisfactorily defined by Σ_1-relationships. In most cases, this will not be sufficient for automatic phasing of the entire data set. In order to proceed further, the following variants have been successfully used:

Symbolic addition. A few reflections (e.g. 4) are chosen which contribute to many triplets, and their unknown phases are assigned "symbols" $a, b, c, d, \ldots$ The set of triplets is then searched for those relationships to which these symbols contribute. As there will in general be far more such relationships than unknown symbols and unknown phases, the most probable value of the symbols and hence of many phases can be determined. This method was first proposed by Zachariasen [33], and was originally used to solve several structures "by hand". It is currently implemented, for example, in the program SIMPEL by H. Schenk [62], which also makes use of quartets. Once the strongest reflections have been determined in this way, the phase set can be extended numerically to the rest of the reflection data using Σ_2-relationships or, in the case of a non-centrosymmetric structure, the tangent formula.

Multisolution methods. In these methods, several reflections are added to the starting set and given arbitrary constant phases. These are then permuted over all possible values. In centrosymmetric structures, each reflection is arbitrarily assigned a sign + or −, so for n such reflections there will be 2^n possible starting sets. For 20 reflections, this is more than 10^6 sets! From each of these starting sets, the phases of the other reflections are determined using triplet relationships. Only in a "correct" or nearly correct starting set will this lead to a result that is relatively free of contradictions. In the non-centrosymmetric case, the phases in the starting set are permuted in steps of 30–50°, which naturally leads to many more starting sets. The calculation and refinement of the other phases is carried out using the tangent formula (eq. 8.17). Since the phases have a higher uncertainty than do those in centrosymmetric ones, non-centrosymmetric structures are in general harder to solve. In particular, the Fourier synthesis at the end of the phase refinement is less likely to show the entire structure.

Various philosophies have been applied to the actual implementation of this process. Originally, when computer time was a more important consideration, a set of 6–12 reflections was carefully chosen, and the phase expansion carried out using further carefully chosen reflections. It is now more common to use a much larger starting set, whose phases are generated randomly. The best known programs which work in this way are probably MULTAN [63], SHELXS [64], and SIR [65].

Figures of merit. Most of the large number of attempts to solve a structure from a random start will fail — a Fourier synthesis based on them will not give a chemically meaningful structure. In order to choose the phase expansions most likely to succeed, a number of *figures of merit (FOM)* have been devised which are calculated during the phase expansion and indicate the relative likelihood that it will lead to a correct solution. The ones implemented in MULTAN are called ABSFOM, Psi(0), and a residual R_α. The value of ABSFOM is calculated from the strongest E-values, and is a measure of the consistency of the triple relationships that have been used. Its value should be 1.0 for a correct solution, and will usually lie in the region of 1.1–1.3.

Psi(0) makes use of a special case of the Sayre equation (Eq. 8.11) using the contributions of the large E-values to a set of very weak reflections. This is particularly useful in centrosymmetric structures, particularly in space group $P\bar{1}$, where the so-called "uranium atom" solution occurs: all phases are positive, giving a large, spurious peak at the origin. In such cases ABSFOM will certainly be large, as all relationships are consistent, but Psi(0) will often indicate which is the correct solution.

The R_α value has a form and value similar to that of the conventional residual factor (see sect. 9.3) which is calculated at the end of a structure refinement. In this case, it shows the difference between the estimated magnitudes of the E-values based on the triplet relationships and their theoretical expectation values. It should be less that about 0.3. In MULTAN, all three values are combined to give a combined figure of merit, CFOM, which should be as large as possible.

In SHELXS, the place of ABSFOM is taken by M_{abs} and an NQUAL value, based in part on low E-values corresponds to Psi(0). The use of negative quartets makes this figure of merit particularly good for eliminating pseudo-solutions. The R_α value is also implemented in SHELXS. The CFOM based on these last two values should be as small as possible.

In some cases, where neither pure direct methods nor pure Patterson methods lead to a solution, a combination of the two may be useful. The program DIRDIF [67], for example, uses direct methods on "difference structure factors". It first locates either a heavy atom or a rigid fragment by a Patterson synthesis. If this does not suffice to give an immediate solution in a difference Fourier, or when direct methods has given only a partial solution which does not develop further, the F_c values for this partial structure (F_p) can be subtracted from F_o to give an approximate F for the rest-structure:

$$F_r = F_o - F_p \tag{8.19}$$

These values, most of which will have some phase information, are then normalized and refined and expanded using triplet relationships or the tangent formula. If successful, the rest-structure is revealed in a subsequent Fourier synthesis. The Patterson-search technique, followed by direct methods is used somewhat differently in the program PATSEE [61].

Problem structures. The available program systems are now so powerful and so convenient to use, and modern computers are so rapid, that most molecular structures seem to solve themselves with little intervention from the crystallographer. When this does not happen, it may be helpful to examine solutions other than the most probable or to increase the number of permutations in the starting set. After that, it is useful to try increasing the number of reflections in the starting set or the number of strong E-values phased by them. Direct methods work best when the space group contains symmetry elements involving translation. In particular, origin problems often occur in space group $P\bar{1}$ (see section 11.5). A trick which sometimes works when a relatively small starting set is being used is to eliminate some of the very strong reflections, or not to use the information from Σ_1-relationships. In most cases where no solution is obtained, the fault lies not in the method employed, but in inaccuracies in the measured data set, or a wrong choice of space group and/or unit cell (see chapter 11).

CHAPTER 9

Structure Refinement

The methods described in Chapter 8 normally produce a *structure solution,* that is a set of atoms defined by coordinates x_i, y_i, z_i for each atom i in the asymmetric unit which describes the crystal structure reasonably well. There are, however, still substantial errors in these parameters, which arise partly from the approximations of the method of solution, partly from errors in evaluating the maxima in the Fourier synthesis, and partly, of course, from errors in the measured data. This leads to the fact that for each reflection hkl, the calculated structure factor F_c or the calculated intensity F_c^2 does not agree exactly with the observed value, so that there is an error Δ_1 or Δ_2:

$$\begin{aligned} \Delta_1 &= \| F_o \mid - \mid F_c \| \\ \Delta_2 &= \mid F_o^2 - F_c^2 \mid \end{aligned} \tag{9.1}$$

These errors correspond to errors *both* in the model *and* in the data. The next stage of the structure analysis, then, is to optimize the values of the parameters so as to make these differences as small as possible. Once the model has been refined in this way, the result may then be called *the* crystal structure — *i. e.* the goal of the X-ray analysis.

9.1 The Method of Least Squares

The method by far the most commonly used to refine a structure is the well known method of *least squares.* It may always be used when a measurable physical quantity Q has a linear dependence on a set of parameters x, y, z in terms of the known values A, B, C, such as:

$$Q_N = A_N x + B_N y + C_N z \tag{9.2}$$

That is, for N values of Q, A, B and C, the same equation, in terms of the same x, y and z, describes the relationship. If the number of observations N is greater than the number of parameters, the parameters are said to be *overdetermined,* and may be evaluated in terms of the measured observations. The "ideal" value for each observation $Q_{N(c)}$ may be calculated in terms of the optimal values of the parameters x, y, z and will, in general, differ from the measured value Q_N by a difference Δ_N.

This is also true, when the values of the parameters are not quite optimal:

$$\begin{aligned} Q_{N(c)} &= Q_{N(o)} + \Delta_N = A_N x + B_N y + C_N z \\ \Delta_N &= A_N x + B_N y + C_N z - Q_{N(o)} \end{aligned} \tag{9.3}$$

The "best" values of the parameters x, y, z are those which give, for all N data, the smallest value to the sum of the squares of the differences, Δ_N^2, i. e. the "least squares". A minimum in $\sum \Delta_N^2$ will occur when a small alteration in the value of any parameter no longer makes an alteration in the value of Q. Put mathematically, this is the situation when the partial derivative of Q with respect to each parameter is equal to 0:

$$\sum_N \Delta_N \frac{\partial Q_{N(c)}}{\partial x} = \sum_N \Delta_N \frac{\partial Q_{N(c)}}{\partial y} = \sum_N \Delta_N \frac{\partial Q_{N(c)}}{\partial z} = 0 \tag{9.4}$$

In a structure determination, the function to be minimized is one of the forms of equation 9.1, relating the observed and calculated values of either the structure factors F or the corrected intensities, F^2:

$$\begin{aligned} \sum_{hkl} w\Delta_1^2 &= \sum_{hkl} w(|F_o| - |F_c|)^2 = \text{min.} \\ \sum_{hkl} w'\Delta_2^2 &= \sum_{hkl} w'\left(F_o^2 - F_c^2\right)^2 = \text{min.} \end{aligned} \tag{9.5}$$

When minimization is based on Δ_1 values, the refinement is said to be "based on F_o," and if on Δ_2 values, to be "based on F_o^2". The differences between the two methods will be discussed further below; the symbol Δ will be taken to refer to both methods. The factor w in equation 9.5 refers to the *weights* given to the various observations so that errors in more accurately measured data "count" more than do those in other data (see section 9.2 below). In order that observed and calculated amplitudes are directly comparable, a *scale factor* must be redetermined after each change:

$$k_1 = \frac{\sum |F_o|}{\sum |F_c|} \qquad k_2 = \frac{\sum F_o^2}{\sum F_c^2} \tag{9.6}$$

This scale factor will not be specifically mentioned in the following equations.

To minimize the sum of the squares of the differences, it is necessary to calculate the partial derivatives of each structure factor with respect to each of the structural parameters, $p_i = x_1, y_1, z_1, U_{11(1)}, U_{22(1)}, U_{33(1)}, U_{23(1)}, U_{13(1)}, U_{12(1)}, x_2, y_2, z_2, U_{11(2)}, U_{22(2)}, \ldots$,

$$\sum_{hkl} w\left(|F_o| - |F_c|\right)\frac{\partial F_c}{\partial p_i} = 0 \tag{9.7}$$

There are thus 9 parameters to refine for each anisotropic atom and 4 for each isotropic atom.

Since the value of F_c does not depend linearly on the parameters p_i, it is necessary to have a starting value for it $F_{c(0)}$ given by the approximate model of the structure, and to work in terms of the small adjustments to the parameters rather than the parameters themselves:

$$\Delta F_c = \frac{\partial F_c}{\partial p_i} \Delta p_i \tag{9.8}$$

In terms of all the parameters, this is:

$$F_c = F_{c(0)} + \frac{\partial F_c}{\partial p_1} \Delta p_1 + \frac{\partial F_c}{\partial p_2} \Delta p_2 + \cdots \frac{\partial F_c}{\partial p_n} \Delta p_n \tag{9.9}$$

So long as the required shifts in the parameters are small relative to the starting values $p_{i(0)}$ (which will only be the case if the model is nearly correct), they can each be represented by Taylor series, e.g. for the first parameter x_1,

$$\begin{aligned} F_c(x_1) &= f \cdot e^{i2\pi h x_1} \\ &= f \cdot e^{i2\pi h (x_{1(0)} + \Delta x_1)} \\ &= f \cdot e^{i2\pi h x_{1(0)}} \cdot e^{i2\pi h \Delta x_1} \\ &= f \cdot e^{i2\pi h x_{1(0)}} \cdot \left[1 + \frac{i2\pi h \Delta x_1}{1!} + \frac{i2\pi h \Delta x_1^2}{2!} \cdots \right] \end{aligned} \tag{9.10}$$

If the Taylor series is then terminated after the second term, what remains is:

$$F_c(x_1) = f \cdot e^{i2\pi h x_{1(0)}} + f \cdot i2\pi h \Delta x_1 \cdot e^{i2\pi h x_{1(0)}} \tag{9.11}$$

It is then easy to take the partial derivatives of these, since the exponential functions are all constant with respect to x_1:

$$\frac{\partial F_c}{\partial x_1} = f \cdot i2\pi h \cdot e^{i2\pi h x_{1(0)}} \tag{9.12}$$

If the minimalization relationship of equation 9.7 is then applied to the F_c values of equation 9.9, the result is:

$$\sum_{hkl} w \left\{ F_o - F_{c(0)} - \frac{\partial F_c}{\partial p_1} \Delta p_1 - \frac{\partial F_c}{\partial p_2} \Delta p_2 - \cdots \frac{\partial F_c}{\partial p_n} \Delta p_n \right\} \frac{\partial F_c}{\partial p_i} = 0 \tag{9.13}$$

Rearrangement and change of sign then gives the *normal equations*, of which there is one for each parameter:

$$\begin{array}{lllll}
\sum_{hkl} w \left(\frac{\partial F_c}{\partial p_1}\right)^2 \Delta p_1 & + \sum_{hkl} w \frac{\partial F_c}{\partial p_1} \frac{\partial F_c}{\partial p_2} \Delta p_2 & \cdots + & \sum_{hkl} w \frac{\partial F_c}{\partial p_1} \frac{\partial F_c}{\partial p_n} \Delta p_n = & \sum_{hkl} w \Delta_1 \frac{\partial F_c}{\partial p_1} \\
\sum_{hkl} w \left(\frac{\partial F_c}{\partial p_2} \frac{\partial F_c}{\partial p_1}\right) \Delta p_1 + & \sum_{hkl} w \left(\frac{\partial F_c}{\partial p_2}\right)^2 \Delta p_2 & \cdots + & \sum_{hkl} w \frac{\partial F_c}{\partial p_2} \frac{\partial F_c}{\partial p_n} \Delta p_n = & \sum_{hkl} w \Delta_1 \frac{\partial F_c}{\partial p_2} \\
\cdots\cdots\cdots & + \cdots\cdots\cdots & \cdots + & \cdots\cdots\cdots = & \cdots\cdots\cdots \\
\sum_{hkl} w \left(\frac{\partial F_c}{\partial p_n} \frac{\partial F_c}{\partial p_1}\right) \Delta p_1 + & \sum_{hkl} w \left(\frac{\partial F_c}{\partial p_n} \frac{\partial F_c}{\partial p_2}\right) \Delta p_2 & \cdots + & \sum_{hkl} w \left(\frac{\partial F_c}{\partial p_n}\right)^2 = & \sum_{hkl} w \Delta_1 \frac{\partial F_c}{\partial p_n}
\end{array}$$

Writing

$$\sum_{hkl} w \frac{\partial F_c}{\partial p_i} \frac{\partial F_c}{\partial p_j} = a_{ij} \quad \text{and} \quad \sum_{hkl} w \Delta_1 \frac{\partial F_c}{\partial p_i} = v_i$$

gives the form

$$\begin{array}{lllllll} a_{11}\Delta p_1 & + & a_{12}\Delta p_2 & \cdots & + & a_{1n}\Delta p_n & = & v_1 \\ a_{21}\Delta p_1 & + & a_{22}\Delta p_2 & \cdots & + & a_{2n}\Delta p_n & = & v_2 \\ \cdots\cdots & & \cdots\cdots & \cdots & & \cdots\cdots & & \cdots \\ a_{n1}\Delta p_1 & + & a_{n2}\Delta p_2 & \cdots & + & a_{nn}\Delta p_n & = & v_n \end{array} \tag{9.14}$$

which may also be written in matrix form, and solved by matrix methods:

$$\begin{pmatrix} a_{11} & a_{12} & \ldots & a_{1n} \\ a_{21} & a_{22} & \ldots & a_{2n} \\ \vdots & \vdots & \vdots & \vdots \\ a_{n1} & a_{n2} & \ldots & a_{nn} \end{pmatrix} \begin{pmatrix} \Delta p_1 \\ \Delta p_2 \\ \vdots \\ \Delta p_n \end{pmatrix} = \begin{pmatrix} v_1 \\ v_2 \\ \vdots \\ v_n \end{pmatrix} \tag{9.15}$$

which can be abbreviated: $\boldsymbol{A}\Delta\boldsymbol{p} = \boldsymbol{v}$.

Multiplication of both sides by the inverse of the matrix $\boldsymbol{A}$, $\boldsymbol{A}^{-1}$ with elements b_{ij} then gives

$$\begin{aligned} \boldsymbol{A}^{-1}\boldsymbol{A}\Delta\boldsymbol{p} &= \boldsymbol{A}^{-1}\boldsymbol{v} \\ \Delta\boldsymbol{p} &= \boldsymbol{A}^{-1}\boldsymbol{v} \end{aligned} \tag{9.16}$$

This allows the necessary parameter shifts to be calculated to improve the model. At the same time, the diagonal elements b_{ii} of the inverse matrix $\boldsymbol{A}^{-1}$ give the standard deviations of these parameters:

$$\sigma(p_i) = \sqrt{\frac{b_{ii}(\sum w\Delta^2)}{m-n}} \tag{9.17}$$

where m is the number of reflections and n is the number of refined parameters.

As a result of the rather crude simplification to a linear relationship in equation 9.8, the results obtained do not represent an exact mathematical solution. Because of this, the process must normally be repeated for several *cycles,* until the parameter shifts Δp_i are small (ideally less than 1%) relative to their standard deviations, when the refinement is said to have *converged.*

Correlation. Occasionally, convergence occurs very slowly or not at all. The refinement may oscillate, with the parameter shifts becoming positive and negative from cycle to cycle without becoming any smaller. On the other hand, the refinement may "explode," the parameter shifts growing exponentially, often being terminated by the computer with an "arithmetic overflow" . The origin of these effects is usually that pairs of parameters are not really independent of one another, but are *correlated.* This means that the effect on the structure factors of a shift in one parameter is virtually identical to some shift of the other. A measure of the extent to which this happens is given by the correlation coefficient, which may have any value between 0 (no correlation) and ± 1 (total correlation for shifts in the same (+) or opposite (−) direction). These may be calculated from the nondiagonal terms of the inverse matrix $\boldsymbol{A}^{-1}$:

$$\varkappa_{ij} = \frac{b_{ij}}{\sqrt{b_{ii}}\sqrt{b_{jj}}} \tag{9.18}$$

In general, these have an absolute value less than 0.5; if they are more than 0.7–0.8, they have a noticeable effect on the refinement, in that they slow it down and increase the standard deviations of those parameters. The cause of significant correlation often lies in incorrect handling of symmetry, especially when the calculation is being carried out in a space group of too low symmetry (see section 11.4).

A common (see, e.g. [34]) example is a structure, properly described by space group $C2/c$, that has been refined in space group Cc. For each pair of atoms, which are related by the 2-fold axis in $C2/c$, only one set of parameters x, y, z needs to be refined, as the coordinates for the second atom are generated from them by the symmetry operation $-x, y, \frac{1}{2} - z$. In Cc, on the other hand, two independent atom positions x_1, y_1, z_1 and x_2, y_2, z_2 must be refined. If these two positions are actually related to one another by symmetry, it will be found that the correlation coefficient for the pairs of parameters $x_1/x_2, y_2/y_2, z_1/z_2$ will be near to ± 1, as it will be for the corresponding displacement factors.

The order of the square matrices is the number of parameters to be refined: for a relatively large structure with e.g. 80 non-hydrogen atoms in the asymmetric unit, with all atoms being refined anisotropically, the order will this be $80 \times 9 = 720$. The building, inversion and multiplication of such large matrices was until recently close to the limit both of storage and time for available computers. Therefore, large structures were not refined with all parameters being allowed to vary in each cycle. Instead, a *block-diagonal matrix* approach was used, in which different parts of the structure were refined sequentially. Today, such approximations are rarely necessary, and further discussion of these methods will not be given here, but see references [9] and [10].

9.1.1 Refinement Based on F_o or F_o^2 Data

As was mentioned earlier, refinement may be carried out either as a function of errors in the structure factors, $\Delta_1 = ||F_o| - |F_c||$ or in those of the intensities, $\Delta_2 = |F_o^2 - F_c^2|$, and in either case, $\sum w\Delta^2$ will be minimized. Until recently, nearly all refinements were carried out "based on F_o," i.e. using Δ_1. For such a refinement, very weak data give problems, since, as a result of counting statistics, for very weak data, the background will occasionally be estimated to be stronger than the peak. This will result in a negative value for F^2, and so, for these data, no value of F_o can be directly calculated. To avoid this problem, it is customary to take an arbitrary value for F_o (e.g. $\sigma(F_o)/4$) for all "unobserved data," say data with $F_o^2 < \sigma(F_o^2)$, so that they can be used for relationships in direct methods. This, of course, introduces a systematic error into the data set. For refinement, these weak data are normally suppressed, by the application of a so-called σ-limit. Since only those data are used which are greater than, say $2\text{–}4\sigma(F_o)$ in this approach, useful information is lost, as the fact that the scattering factors for the atoms in the unit cell add vectorially to zero for some data is in principle as significant as the fact that they add to a large absolute value for other data.

This problem is less serious when F_o^2 data are used directly in refinement, and $\sum w\Delta^2 = \sum w(F_o^2 - F_c^2)^2$ is minimized. In this case, *all* measured data, including those with negative values of F_o^2 may be contributed to the refinement. Experience has shown that for good data, with few weak reflections, both methods lead to very much the same results. For weak data sets, and those with special problems such as superstructures (see chapter 11) refinement based on F_o^2 is definitely preferable, and yields positional parameters with standard deviations some 10–50 % less than those obtained from the same data refined against F_o. In addition, the atomic displacement parameters as well as the derived bond lengths and angles tend to be more chemically reasonable. For these reasons, refinement based on F_o^2 is now generally preferred. In particular, it has been implemented in the very widely used program SHELXL [68], the use of which is illustrated with a practical example in Chapter 15.

In addition to the normal linear least-squares refinement which has been described here, various non-linear methods have been proposed. A quite different approach to refinement is based on "maximum entropy". Further information about such methods is given in International Tables C, chapter 8.2.

9.2 Weights

In Eq. 9.5, a weight w is applied to both Δ_1- and Δ_2-values, allowing for the fact that the various reflections making up the data set have not all been measured with the same accuracy. The most important factor contributing to the error in a measured intensity F^2, and in the structure factor F_o derived from it, is the statistical error σ from the counting statistics of the diffractometer measurement (see section 7.4.2). Relatively, it is larger for weak reflections than for strong ones. In many cases, in particular when refinement is based on F_o, it suffices to give each reflection a weight of the form:

$$w = 1/\sigma^2 \tag{9.19}$$

In refinements based on F_o, $\sigma(F_o)$ is used, while in those based on F_o^2, $\sigma(F_o^2)$ is required.

Unfortunately, it is not only statistical errors which affect the measured values of the intensities; there are other, more or less systematic errors. Among the most important are ignored or badly corrected effects of *absorption* (see section 7.4.3) or *extinction* (section 10.3). These (particularly the latter) especially affect strong intensities at low diffraction angle. It is, of course, much the best procedure, if possible, to make accurate corrections for such factors. Since this is often not possible, it is usual to lower the weights based on counting statistics, which are often very high, for the data most likely to be affected. This is especially important for refinement based on F_o^2 values, since the double squaring (equation 9.5) makes it very sensitive to errors in $|F_o| - |F_c|$.

A simple weighting function which will allow for this is:

$$w = 1/(\sigma^2 + kF_o^2) \qquad (9.20)$$

The factor k is usually determined empirically, and typically has a value of 0.001–0.2. It can be refined as a variable parameter, but this is often unstable, and it may better be optimized through variation.

Much more complex weighting schemes have been proposed, which have further adjustable parameters. The objective of these is to ensure that the sum of the weighted squares of the errors (the variances) in any class of reflections is the same as that in any other. In particular, this should be so for the weakest, the intermediate and the strongest data. In the previously mentioned program SHELXL, the function most commonly used is

$$w = 1/\left(\sigma^2(F_o^2) + (a \cdot P)^2 + b \cdot P\right) \qquad \left(P = \frac{1}{3}\max\left(0, F_o^2\right) + \frac{2}{3}F_c^2\right) \qquad (9.21)$$

in which the parameters a and b are chosen to minimize the differences in the variances for reflections in different ranges of intensity and diffraction angle. Such sophisticated weighting schemes should be applied in the last stages of refinement only.

Particular care is required for the refinement in space groups of high symmetry. In the most symmetric Laue group $m\bar{3}m$, for example, normally several equivalent reflections will be measured for each independent datum (N may have values 2, 4, 6, 8 ... 48). When these data are averaged, and the statistically correct standard deviation calculated, it may become very small, since statistical errors have now virtually vanished with respect to systematic ones. What is important is that the difference between the errors of one class of reflections and another may have become very great. A reflection of the type h00 will be at most one of six symmetry equivalents: $\pm h00$, $0 \pm k0$, $00 \pm l$. A general reflection hkl in this Laue group will be one of 48 equivalents: $\pm h \pm k \pm l, \pm h \pm l \pm k, \pm k \pm h \pm l, \pm k \pm l \pm h, \pm l \pm h \pm k, \pm l \pm k \pm h$. If these data are all weighted by $w = 1/\sigma^2$, then the second class will clearly get a higher weight, although in fact the errors for reflections of equal intensity are likely to be evenly distributed between the two classes. To avoid this, it is useful either to use the standard deviations of the individual data, or to use unit weights ($w = 1$).

9.3 Crystallographic *R*-Values

In order to indicate how well a structural model actually conforms to "reality," so-called "residuals" or "R-factors" are evaluated. The "conventional" R-factor:

$$R = \frac{\sum_{hkl} \Delta_1}{\sum_{hkl} |F_o|} = \frac{\sum_{hkl} || F_o | - | F_c ||}{\sum_{hkl} | F_o |} \tag{9.22}$$

when multiplied by 100 % gives the average relative deviation between the observed and calculated structure factors as a percent. This is always quoted in the literature, even when the refinement was not based on F_o. It is important to notice which selection of data has been used for the calculation of the conventional R-factor (e.g. those with $F_o > 3\sigma(F_o)$). No allowance is made for the weights which were used in the refinement. If weights are taken into consideration, the resulting R-factor is usually worse (i.e. larger). The *weighted R-factor* is directly related to the quantity that is minimized in the least-squares refinement. Its changes show whether changes in the structure model are actually meaningful. Unfortunately differences in weighting schemes make it difficult to use for comparing one structure with another. It is defined differently, depending on whether the refinement is based on F_o, or on F_o^2:

$$wR = \sqrt{\frac{\sum_{hkl} w\Delta_1^2}{\sum_{hkl} wF_o^2}} \tag{9.23}$$

$$wR_2 = \sqrt{\frac{\sum_{hkl} w\Delta_2^2}{\sum_{hkl} w(F_o^2)^2}} = \sqrt{\frac{\sum_{hkl} w(F_o^2 - F_c^2)^2}{\sum_{hkl} w(F_o^2)^2}} \tag{9.24}$$

Different authors define R-factors differently. In this book, R or wR implies a factor based on F_o, while the subscript 2, e.g. wR_2, implies one based on F_o^2.

Because the terms in equation 9.24 are squared, wR_2 values are generally twice or three times as large as are wR values for refinements based on data of comparable quality. They are much more sensitive to small errors in the structure model, such as disorder or missing H-atoms. A further index used to indicate the quality of a refinement is the "goodness of fit," S:

$$S = \sqrt{\frac{\sum_{hkl} w\Delta^2}{m - n}} \tag{9.25}$$

(m = number of reflections, n = number of parameters). The difference $m-n$ gives the overdetermination of the structure. For a correct structure with a suitable weighting scheme, S will have a value close to 1. It is important to specify whether refinement was carried out using F_o or F_o^2.

For a good data set and a structure without special problems, the refined structure should give a wR_2-value less than 0.15 and wR and R values less than 0.05. A completely random non-centrosymmetric structure will give an R-value of about 0.59; in a centrosymmetric space group, this value rises to 0.83. The actual minimum R-value which can be obtained for a correct structure depends both on the quality of the data and on the suitability of the structure model. Since the atom formfactors are calculated for spherical atoms, bonding electrons and lone pairs are not considered by the model. In addition, the assumption of harmonic vibrations which are implied by anisotropic atom displacement parameters is not strictly correct. In the case of H-atoms, both of these problems arise. The single electron is significantly shifted along the direction of the bond (see section 9.4), and it is never possible to refine hydrogen atoms anisotropically, even in peripheral methyl groups where vibration is highly anisotropic. Even with good data, somewhat larger than usual R-values must be expected for those structures in which a large proportion of the electron density is localized in bonds, H-atoms and lone pairs.

9.4 Refinement Techniques

Initial refinement strategy. As soon as a plausible structure model has been found by Patterson or direct methods, *before* calculating a difference Fourier synthesis, it is sensible to put it through a few cycles of least squares refinement in order to improve the positions of the known atoms and, by inspection of the resulting R-factors, to decide whether it is worthwhile to proceed further. As a general rule, a difference Fourier is likely to be useful when the conventional R-value is 0.4 or less. The wR_2-value may well lie in the range of 0.5–0.7. If there is uncertainty about the types of atoms present, or particularly if the structure is entirely unknown, it may help first to refine only the scale factor (see equation 9.6). In subsequent refinement cycles, isotropic atomic displacement parameters U may also be refined, starting from an initial value of, say, 0.01 Å^2 for heavy atoms to 0.05 Å^2 for C-atoms. From the behavior of these, it is often easier to see that atoms have been assigned the wrong type or that they are not there at all than it is from difference Fourier maps. If there is in fact *no* atom on the site chosen, the displacement parameters and their standard errors will rise rapidly from cycle to cycle, in order to lower the electron density at the site by modelling it as extremely large vibrations. If the displacement parameter does converge, but to a value that is unreasonably high, this may indicate that the position is actually occupied by a lighter atom (e.g. C in place of N). Similarly, an unreasonably low or even negative displacement parameter indicates that the site is actually occupied by a heavier atom (e.g. N or O in place of C).

It is also possible that a model with several false atoms in it may end up with the correct atoms having low or negative U-values, in order to give these atoms more weight than the false ones.

Completion of the model structure. A difference Fourier synthesis based on the improved model should lead to an essentially complete structure with a wR_2-value of about 0.3 and an R-value less than 0.15. Anisotropic displacement parameters can now be introduced for all non-hydrogen atoms (except possibly other very light atoms, such as Li and B) although for large structures, this is best done in stages, beginning with the heaviest atoms. Many computer programs will automatically convert isotropic U-values to the appropriate anisotropic ones. Otherwise, it is usually satisfactory to take $U_{11} = U_{22} = U_{33} = U_{iso}$ and to set the "mixed" terms to zero (for trigonal or hexagonal crystals $U_{12} = U_{iso}/2$ and for monoclinic crystals $U_{13} = -U_{iso} \cos \beta$). At this stage it is appropriate to introduce a weighting scheme or to examine the suitability of the initial, default scheme. If possible, H-atoms should now be located in a difference map (see below). A complete structure model should refine to give a wR_2-value less than 0.15 and an R-value below 0.05, possibly after some of the corrections mentioned in Chapter 10. Values higher than these are only acceptable if the cause is known. There are occasionally situations in which an incorrect structure refines into a *pseudominimum.* Because of this, it is always necessary to examine the refinement critically, and to make certain that the final model makes sense in terms of the known chemical and physical properties (see also Chapter 11).

9.4.1 Location and Treatment of Hydrogen Atoms

Depending on the quality of the data set, the presence or absence of heavy atoms and the extent of thermal motion of all the atoms, hydrogen atoms may in some cases be found and refined with isotropic displacement parameters, while in others they may not be located at all. A single H-atom has very little scattering power, but in many cases, their sheer number makes a contribution of 10–20 % of the electron density. This is particularly so when protecting groups such as t-butyl or trimethylsilyl are present. Hydrogen atoms must whenever possible be included in the structure model, as otherwise the quality of the entire solution will suffer significantly.

In order to obtain an optimal difference Fourier map, it is essential that the partial model be as well refined as possible, and that all necessary corrections – especially for absorption (section 7.4.3) and extinction (section 10.3) – have been made to the data. As indicated above, these corrections are greatest for reflections at low scattering angle, which also contain most of the information about the H-atoms. It can be advantageous to refine the structure using only "high-angle" data, e. g. those with $\vartheta >$ 25° with Cu-radiation. These data mainly represent core electrons of heavier elements and are less affected by absorption and extinction. When the positions obtained from this refinement are used to phase a difference Fourier synthesis based on *all* the data, the chance of locating hydrogen atoms is maximized. An example of the location of H-atoms in a difference synthesis has already been given in Fig. 8.1 (sect. 8.1).

There are some cases in which H-atoms either cannot be located or once located do not refine well. In other cases, refinement is undesirable as it makes the data : parameter ratio too small. In all of these cases, the positions of the H-atoms are, if possible, calculated in terms of the expected geometry of their environment and then refined together with the atoms to which they are bonded as *rigid groups* (see below) or as *riding atoms*. This means that as a C-atom refines, the H atoms bonded to it are shifted in parallel so that their geometry is unchanged.

The bond lengths to H-atoms determined by X-ray diffraction are significantly shorter than are those determined by other methods, e. g. neutron diffraction, which gives true internuclear distances. Examples are C—H 0.96 instead of 1.08 Å, N—H 0.90 Å and O—H as little as 0.8–0.85 Å. The cause of this is the fact that X-rays measure the distances between centers of electron density, and the single electron of the H-atom is naturally displaced along the direction of the bond. This effect must be borne in mind when bond lengths are discussed, especially in hydrogen-bonded structures X—H· · ·Y where the shortening of the X—H distance will lengthen the H· · ·Y distance.

In general, only isotropic atomic displacement parameters are used for H-atoms. When the data are sufficiently good, the data:parameter ratio sufficiently high, and when there are no heavy atoms in the structure, they may be refined individually. If there is a need to "save" parameters or if free refinement is impossible, the displacement parameters may be refined in groups, *e. g.* one value for the three hydrogens of a methyl group or the five of a phenyl group. Finally, these parameters can be fixed or constrained to be some factor times the values for the atoms to which they are bonded. Experience indicates that a factor of 1.2–1.3 is often satisfactory. At the end of the refinement – particularly if calculated bond lengths have been used to position the H-atoms – it is wise to refine the hydrogen displacement parameters at least in groups for a cycle. This may indicate false positions, as the values of U will rise to something above 0.15 Å^2 with high estimated errors. These errors often occur, for example, when methyl groups bonded to aromatic systems may have two alternative rotational orientations with similar energies.

9.4.2 Restricted Refinement

In some structures, the shifting of individual atoms has little effect on the calculated structure factors and consequently little effect on the minimization of the R-values. This problem becomes apparent when the parameters do not become stable after a few cycles but give constant or oscillating shifts and positions corresponding to chemically unreasonable bonds and angles. It occurs especially for weakly diffracting H-atoms in structures with strongly vibrating groups such as water of crystallization or other loosely held solvent molecules. Such atoms make little contribution to the scattering. Another cause of unstable refinement is disordered groups (see sect. 10.1), also found particularly commonly in solvent molecules. Sometimes two alternative configurations of a molecule are statistically distributed over a particular site in the unit cell. In such a case, all the parameters cannot be refined independently as those for atoms which overlap in the averaged structure will be strongly correlated and lead to poor convergence. It is then appropriate to seek to refine the structure using geometrical knowledge.

Constrained refinement. When a structure contains moieties whose geometries are well known, such as phenyl rings, these can be refined as *rigid groups* with fixed bonds and angles. Instead of refining three positional parameters x, y, z for each atom, only those for a single "pivot atom" are refined together with three orientation parameters $\varphi_x, \varphi_y, \varphi_z$ defining the rotation of the group about the three axial directions. This is a great reduction on the number of parameters, especially if a single isotropic displacement parameter is used instead of one or six for each atom. This possibility is particularly attractive when a data set is very weak and the number of significant reflections is small and/or the structure is very large and the data:parameter ratio too small. Ideally, this ratio should be greater than 10 : 1 or at least 7 : 1.

Restrained refinement. In this approach, every atom is refined as usual, but expected bond lengths and sometimes angles are given (with appropriate weights) as additional data. In effect, this means that differences between observed and calculated interatomic distances are contributed to the least squares sums in the same way as are the structure factor differences (equation 9.5). There is thus no reduction in the number of parameters, but there is an increase in the number of "observations". In contrast to the *constraints* given above, these *restraints* can be made weak by decreasing the weights assigned to them, or effectively assigning a standard deviation within which they may vary.

Refinement of macromolecular structures. In macromolecular (mainly protein) structures, the reflection/parameter ratio is normally too small to allow for refinement of individual atomic parameters. For this reason, and because the geometry of amino acid residues is well known, restraints and constraints are almost invariably used. Other refinement techniques besides least squares are also used, in particular application of maximum likelihood, molecular dynamics and simulated annealing. For further information, see references [54, 55].

9.4.3 Damping

When a structure is difficult to refine, particularly when the structure model is far from optimal, oscillation and divergences can often be controlled by a "damped" refinement. This effect is accomplished by different programs in different ways. In some, all shifts are multiplied by a constant factor; in others a more complex mathematical method is used. Such techniques may lead to an underestimation of the errors of the parameters. At least in the last cycle of refinement, there should be no damping in order to obtain reasonable estimates for these errors. If this is impossible, there is a high probability that the space group is wrong or that some other error has occurred (see Chapter 11).

9.4.4 Symmetry Restrictions

The symmetry elements of the unit cell reduce the number of atoms to be refined to those of the asymmetric unit. In some cases, atoms lie on special positions (see

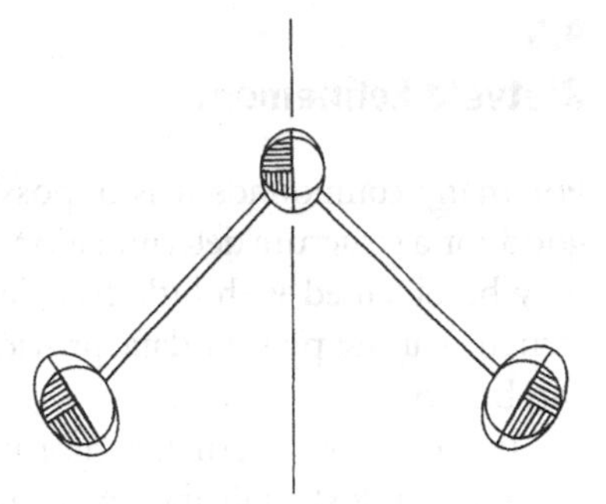

Fig. 9.1. Symmetry restriction to displacement ellipsoids – in this case a mirror plane.

sect. 6.4) e.g. at $x, \frac{1}{4}, z$ on a mirror plane normal to $\boldsymbol{b}$. The *special parameter* (here $y = \frac{1}{4}$) must not be refined. A more complicated situation occurs in space groups of higher symmetry, where positional parameters may be coupled. The special position on a 3-fold axis parallel to the body diagonal of a cubic cell (e.g. the position 32f (x, x, x) in $Fm\bar{3}m$, № 225) implies that $x = y = z$ and thus that these three parameters must always shift together. When an atom is on a special position, its displacement ellipsoid must also conform to the point symmetry of that position, and this usually places restriction on anisotropic atomic displacement parameters. For example, an atom in the monoclinic space group $C2/c$ on position 4e $(0, y, \frac{1}{4})$ lies on a 2-fold axis parallel to $\boldsymbol{b}$, so one of the principal axes of the ellipsoid must lie in this direction and the other two must be normal to it. The implication of this is that the cross terms with a y-component (U_{12} and U_{23}) must be zero and thus must not be refined. A summary of the restrictions for all positions in all space groups is given in [35] or in *Int. Tab. C*, table 8.3.1.1. Care must be taken that the atom chosen lies on a position corresponding in form to the first symmetry position in the space group table. Figure 9.1 shows an atom doubled by a mirror plane. By the symmetry operation the orientation in the space of the displacement ellipsoid is altered. It is thus important to remember that the U_{ij} terms for such an ellipsoid will change when a symmetry equivalent ellipsoid is drawn. Most programs today automatically allow for restrictions on the positional and displacement parameters of atoms in special positions.

9.4.5 Residual Electron Density

At the end of a successful structure refinement, a final difference Fourier synthesis must be calculated, and it should show no significant electron density peaks or troughs. For light atom structures, the maximum should lie in the range ± 0.1–$\pm 0.3 e\text{Å}^{-3}$. For heavy atom compounds, it is normal to find ripples of up to $\pm 10\,\%$ of the atomic number of the heavy atom per Å^3 at a distance of 0.6–1.2 Å from it.

9.5 Rietveld Refinement

For many compounds it is impossible to grow single crystals that are sufficiently good for a structure determination, although very good powder diffraction patterns may be obtained with little trouble. In such cases, it is often possible to refine the structure using powder data, providing a sufficiently good starting structural model can be devised.

In a powder pattern, the information contained in the spatial distribution of the reflections is lost. Only the single dimension of the scattering angle remains, and this makes the problem of indexing (assigning each measured reflection to specific lattice planes hkl) and specific intensity measurement difficult or even impossible.

The Rietveld method of structural refinement is an elegant solution to this problem, because the refinement is made, not against individual structure factors for each reflection, but against the entire profile, taken in small steps of the scattering angle, and considering all the reflections which contribute to each step in turn. It is thus necessary to measure the entire profile accurately.

The method was originally designed for the results of neutron diffraction by powders, where each reflection gives a "line shape" which makes a Gaussian contribution to the overall profile. X-ray diffraction using counter diffractomenters, on the other hand, gives much more complex line shapes, and various algorithms have been designed to describe them. The most widely used functions are the pseudo-Voigt- and Pearson-VII-functions. Refinement must include the unit cell parameters and appropriate profile shape parameters as well as the atomic positional parameters and (usually) isotropic atomic displacement parameters. The constraints and restraints mentioned in section 9.4.2 may often be very suitable applied to the Rietveld method. Powder diffraction data collected from flat specimens often show a "preferred orientation effect". This occurs when the crystallites in the powder have a strongly anisotropic form, *e.g.* needles or plates. Such crystallites will tend to lie parallel to the plane of the sample holder, thus greatly enhancing the diffraction by those lattice planes that are roughly parallel to the needle axis or plate face. A parameter to correct for preferred orientation can be included in the refinement process.

The quality of a Rietveld refinement may be inferred by the calculation of an "R-factor" similar to that used in single-crystal refinement:

$$R_{wp} = \sqrt{\frac{\sum w_i (y_{i(o)} - y_{i(c)})^2}{\sum w_i (y_{i(o)})^2}}$$

The quantity minimized is the sum of the squares of the weighted differences between the measured ($y_{i(o)}$) and calculated ($y_{i(c)}$) values for each measured point on the powder diagram that is used for least-squares refinement. In contrast, the "Bragg-R-value" is calculated on the basis of the individual n reflections. This calculation is sensitive to the resolution of overlapping reflections, which unfortunately cannot be done without reference to the refined model.

$$R_{\mathrm{B}} = \frac{\sum |I_{n(\mathrm{o})} - I_{n(\mathrm{c})}|}{\sum I_{n(\mathrm{o})}}$$

The quality of a refinement is best shown using a figure on which the observed, calculated and difference powder diagrams are all shown (see Fig. 9.2)

This method is particularly appropriate for the refining of small structures of known structure type — e.g. when the new material is isostructural with one whose structure has already been determined by single crystal methods, and the positions of the atoms in the new structure only differ slightly from those in the known structure. The *determination* of a structure *ab initio* from powder data is difficult and up to now has only been carried out in a few hundred cases. For it to be successful, the following steps are required:

- The unit cell of the unknown structure must be determined by indexing the powder pattern. This is often difficult, particularly for monoclinic and triclinic crystals.
- Deconvolution of the pattern must yield enough non-overlapping intensities which can be indexed unambiguously for use in Patterson or direct methods. This is critical and usually only 50–200 such reflections are to be found.
- The space group must also be determined from the data. One difficulty is that the inevitable precise coincidence of some possibly symmetry unrelated data makes Laue groups in the same crystal system (e.g. $4/m$ and $4/mmm$) impossible to distinguish.

Since powder methods are not within the scope of this book, the reader is referred to relevant literature [24] for further details.

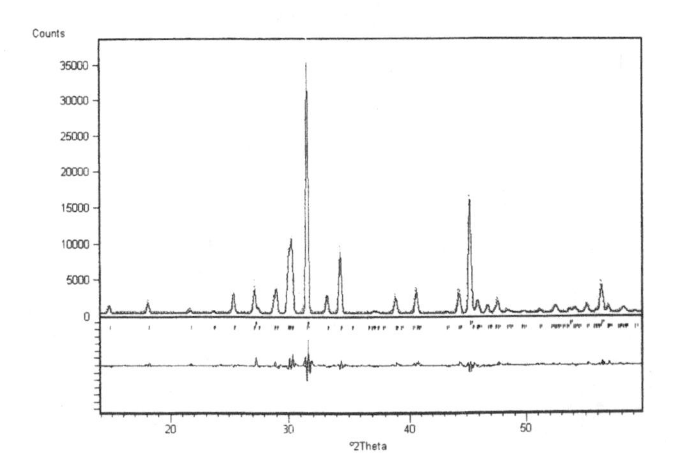

Fig. 9.2. Example of a Rietveld refinement. Dots: experimental data; solid line: calculated pattern; below: difference pattern.

CHAPTER 10

Additional Topics

10.1
Disorder

There is no sharp boundary between substances which are crystals with long range order in three dimensions and those which are amorphous solids with no long range order. When long range order characterizes most of a substance, and only small parts are disordered, the structure can usually be solved and refined by conventional methods. It is only because of very large, anisotropic displacement parameters and perhaps chemically unreasonable arrangements of atoms that the *disorder* makes itself apparent. In such cases, an exception must be made to the rule of order in three dimensions, and the model changed to allow a disordered section. The most common types of disorder are described in the following sections:

10.1.1
Site Occupancy Disorder

Even when it is possible to describe a structure completely in terms of the occupation of sites in the unit cell, some of these sites may be occupied randomly by *different* types of atoms. This is a very common situation in mineral structures, where a site is occupied by a statistical array of two or more atom or ion types of similar size. In zeolites, for example, the Si- and Al-atoms are often statistically distributed over the tetrahedral sites in the three dimensional aluminosilicate framework. This sort of disorder is widespread, e.g. in alloys and other mixed crystals.

In the structure model, such a situation is treated by assigning the different atoms the same site x, y, z and the same displacement parameters U_{ij}. One of the atoms is then given a site occupancy of k and the other of $1 - k$. There may be a strong correlation between the site occupancy and the displacement factor, in which case it is best to refine them in alternating cycles of refinement until they converge.

A common special case of this type of disorder is that of partial occupancy, when a single atom is disordered with a vacancy. In such cases, the site occupancy may be varied freely – if necessary with a fixed displacement factor. A typical example is that of the tungsten bronzes A_xWO_3, where A is an alkali metal. These structures consist of a framework of WO_6 octahedra containing channels which are partially occupied by alkali metal ions. In zeolites and many other compounds, including some molecular crystals, partial occupancy of sites by water or other solvent molecules is common.

There are also compounds which, despite having an exact chemical composition still show partial site occupancies. For example, nearly all cubic pyrochlore structures of the $RbNiCrF_6$ type with formulae $A^{I}M^{II}M^{III}F_6$ (A = alkali metal) have the divalent and trivalent cations distributed over a *single* set of positions, even when the atomic radii are substantially different.

10.1.2 Positional and Orientational Disorder

Positional disorder occurs when an atom, a group of atoms or an entire molecule is statistically distributed over two or more positions. *Orientational* disorder implies that a molecule is distributed over two orientations, usually related by a rotation, reflection or inversion, and usually sharing a common center of gravity. This is particularly common in structures which contain a nearly spherical group of atoms, e.g. NH_4^+ cations, ClO_4^-, BF_4^- or PF_6^- anions or CCl_4 molecules (Fig. 10.1).

Frequently such ions are found on sites of point symmetries inconsistent with their own, e.g. a tetrahedral molecule on an inversion center, so that the alternative orientation is generated by a symmetry operation. In such cases, it is important to be certain that the "disorder" does not arise from the incorrect choice of a space group of too high symmetry.

If different, energetically similar conformations of a molecule can pack equally well into a crystal structure, this can also lead to positional disorder. Examples of this are the two alternative conformations of a methyl group bonded to an aromatic ring, or the more widespread disorder resulting from different conformations possible for large rings such as crown ethers. Metal complexes containing π-bonded cyclopentadienyl rings often exhibit disorder resulting from alternative positions due to rotations of the rings about their axes.

The split atom model. When two orientations of a group result in atom positions which are at least about 0.8 Å apart, separate maxima may usually be found in the Fourier syntheses, and the disorder may be refined using the so-called *split-atom*

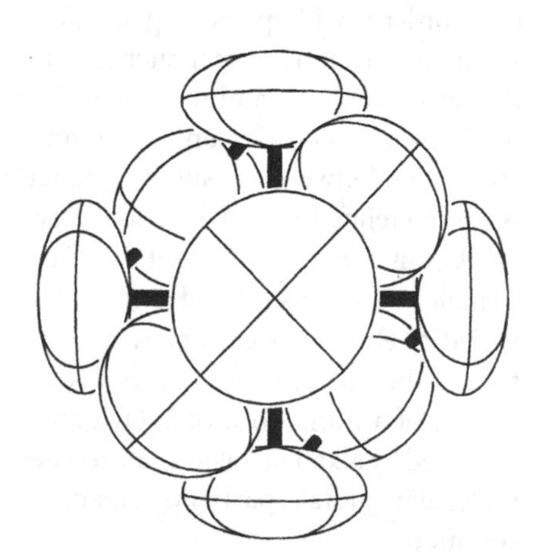

Fig. 10.1. Representation of a disordered BF_4-anion using 10 partially occupied F-sites.

model. The two partial atoms are refined independently, except that the sum of their site occupancies is constrained to be 1. If the two atoms are much closer to one another, their electron densities will seriously overlap, and the refinement, particularly of their displacement parameters, will suffer greatly from correlation. In this case, it is useful to constrain the displacement parameters to be the same but to refine the positional parameters separately. If this is successful, it may be possible then to hold the positional parameters constant and refine the temperature factors separately. In fact, in most such cases, the original difference Fourier synthesis showed only *one* maximum, at the midpoint of the two positions, and refinement of it results in grossly anisotropic displacement parameters corresponding to a "cigar-shaped" ellipsoid. Whenever one of the principal axes of the displacement ellipsoid exceeds 0.2–0.3 Å^2, the possibility of splitting the atom into two parts should be considered. It is a particularly time-consuming, intuitive process to select the most suitable disorder model in such cases. On the one hand, it is important to refine the displacement parameters anisotropically, since it is precisely the atoms at the edge of a molecule which are most subject to large vibrations (see below). On the other hand, in disorder cases, these displacement parameters are highly correlated and therefore refine poorly. If the geometry of the group is well-known, it may be possible to apply suitable restraints or constraints to aid the refinement (see section 9.4). The proper handling of such disorder is important, even when the part of the structure affected is itself unimportant, because every fault in the structure model detracts from the quality of the refinement, and this will affect the results for the "interesting" part of the structure as well.

Dynamic or static disorder? Thus far, it has been assumed that a disordered group is statistically distributed among two or more "rest"-positions, about which they vibrate. It is easy to see that with sufficient thermal energy, it will be possible for the atom(s) to pass back and forth between the alternative positions, giving a dynamic equilibrium. In the example of the cyclopentadienyl complexes, this would consist of the free rotation of the Cp-rings about the ring-metal axis. In such cases, a ring-shaped cloud of electron density may be modelled as well by a split-atom model with displacement parameters that are highly anisotropic in the ring plane as for a true positional disorder. In a single data set, X-rays cannot distinguish between dynamic and static disorder. It may be possible, by carrying out investigations at different and especially at much lower temperatures to observe changes in the displacement parameters which can be interpreted. By this means, it may be possible to observe a freely rotating group freeze into a single position. On the other hand, a dynamic situation may pass smoothly over into one of static disorder, and there will be no significant change in the diffraction. To settle this matter, other physical methods, such as thermal analysis should be undertaken. A third possibility is that as the crystal is cooled, the dynamic disorder may be transformed to an ordered phase, but that two alternative ordered orientations exist. If the ordering occurs in different ways in different regions of *domains* of the crystal, whose size is greater than the coherence length of the X-rays (ca. 100–200 Å), the result is called *twinning*, and is discussed further in the next section. If the domains are smaller than this, the result is effectively static disorder, where the orientations vary randomly from one unit cell to the next.

All crystals are subject to the inevitable property that as the temperature increases, an increasing number of lattice sites will become unoccupied, and the missing atom will migrate either to the surface (Schottky defect) or to an interstitial site (Frenkel defect). However, the concentration of such sites is so small (about 0.01 % at normal temperatures), which is too small to have a measurable effect on F_c-values, and so may be ignored in crystal structure analysis.

10.1.3 One- and Two-Dimensional Disorder

Layer structures exist which have excellent ordering in the layers, but where the third dimension, the stacking direction, is disordered. This *one-dimensional disorder* is visible in the diffraction pattern, e.g. on a film, because long streaks occur in the direction of the stacking instead of sharp reflections (Fig. 10.2).

Despite this streaking, the positions where the reflections expected in a fully ordered structure occur do indeed show maxima on the streaks. Because of this, such maxima, usually with high backgrounds, will register as peaks on a serial diffractometer, and without film or area detector observations, the disorder may be missed. The result of this will be a wrong structure, in that it will consist of a superposition of the different orientations of the layers. Such structures can only be solved properly by a full mathematical treatment of the intensity distribution along the streaks (see, e.g., the structure determination of $MoCl_4$, [49]).

Occasionally, chain structures are encountered, which are only ordered in *one* dimension. These structures are said to have *two-dimensional disorder.* Photographs

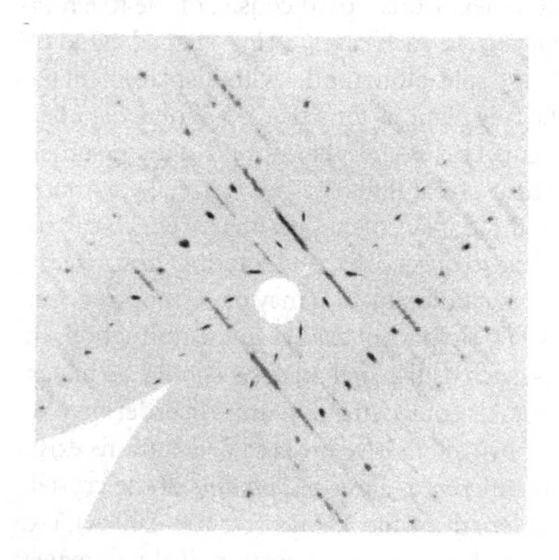

Fig. 10.2. Section of a reciprocal lattice plane (recorded using an area detector) with one-dimensional disorder.

of their diffraction patterns show diffuse blackening within the reciprocal planes normal to the axis of the chain.

10.1.4 Modulated Structures

During the last twenty years a series of structures has been found showing a new special type of long range ordering: While a main part of the structure can be described with the usual 3D ordering of a normal translational lattice, another part of the structure shows continuous variation of atom positions, and sometimes of site occupations or displacement factors, from unit cell to unit cell. This can be described by a periodic function, usually a sine function. When the wavelength of this function is a rational small multiple of the translation period of the basic unit cell (commensurate modulation) the effect can be described as a "superstructure" with a correspondingly enlarged super cell (Fig. 10.3a). If the ratio is irrational, the modulation is called incommensurate. Such a case may be recognized in the diffraction pattern by the occurrence of "satellite reflections" wich are grouped around the "main reflections" of the basic structure (Fig. 10.3b).

Modulations can occur in 1, 2 or 3 dimensions, and indexing each resulting satellite reflection will require 1, 2 or 3 so-called "q-vectors" to relate it to the corresponding normal reflection. Thus, a total of 4–6 indices are required for each reflection. The proper description of these structures thus requires the use of 4-, 5- or 6-dimensional space groups (see, e.g. [5] and Int. Tab. C, section 9.8). Modulated structures can be refined using the program JANA [79].

10.1.5 Quasicrystals

A further way in which normal three-dimensional translation symmetry can be broken occurs in what have become known as *quasicrystals*, and are exemplified by alloys occurring e.g. in the Al/Mn system or in $TaTe_{1.6}$. In these systems, the diffraction patterns clearly show crystallographically "forbidden" symmetry, such as 5-, 8-, 10- and

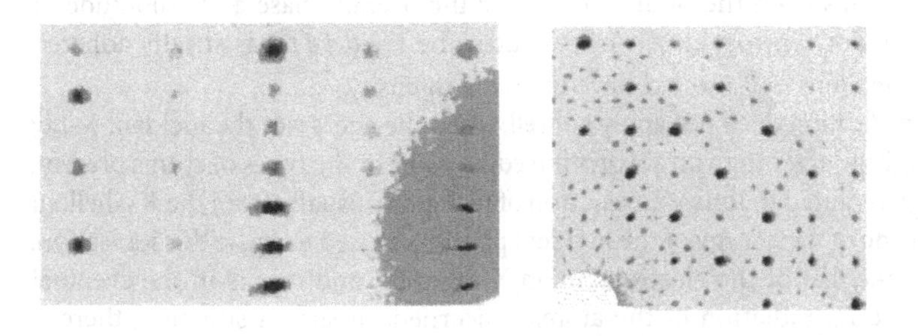

Fig. 10.3. Section of a reciprocal lattice plane (recorded using an area detector) with **a)** superstructure reflections in the horizontal direction, and **b)** satellite reflections typical of an incommensurate modulated structure.

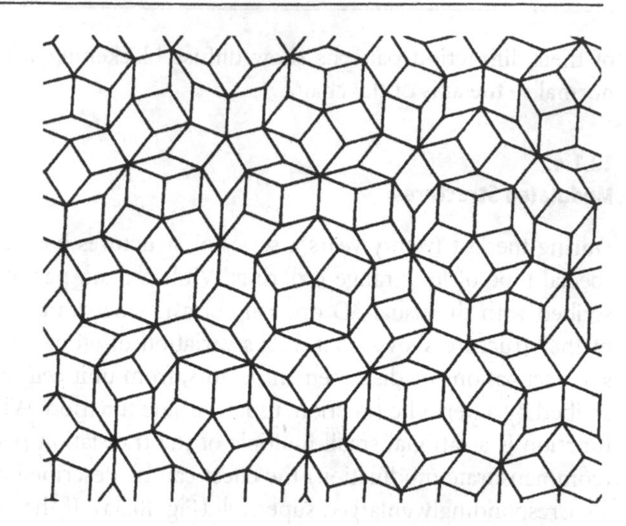

Fig. 10.4. Fivefold "Penrose"-tiling.

12-fold axes. Such structures cannot be explained in terms of a simple lattice, but it is necessary to describe it by space-filling packings of, for example, two different rhombohedra *without* three-dimensional periodicity but able to adopt icosohedral ($m\bar{3}5$ or I_h) symmetry. It has so far been possible only to understand the principles of this unusual diffraction phenomenon, not to solve a structure completely. A two-dimensional analogue of quasi-crystals is found in the well-known *Penrose-tiling* (Fig. 10.4) a two dimensional aperiodic array with two different shapes of tile.

10.2 Anomalous Dispersion and "Absolute Structure"

So far in this book, the calculation of structure factors F_c for a structure model has assumed classical, elastic scattering of X-rays, in which the scattering atoms alter the radiation neither in amplitude nor in phase (except for the phase shift of π implicit in the scattering of electromagnetic waves). The calculation has been based on atomic scattering factors f_i whose vectorial addition takes into consideration the displacements of the atoms from the origin to evaluate the overall phase and amplitude of each reflection. For centrosymmetric structures, the *Friedel Law* is strictly adhered to, and the reflections hkl and $\bar{h}\bar{k}\bar{l}$ have the same intensity.

This simplification does not apply so well, when the energy of the incident X-ray photons is slightly larger than an absorption edge of one of the types of atoms present, i.e. the energy required to ionize an electron of that atom, usually from the K-shell. In such cases, some of the energy of the incident photons is used to cause this ionisation. The effect is like that of the electrons in an X-ray tube, and results in the eventual emission of the K_α radiation of the atoms concerned. In such a situation, there is naturally an increase in background radiation. The photons which are not absorbed in this way do nonetheless interact particularly strongly with these atoms, and are altered slightly both in magnitude and in phase, and this effect is called *anomalous scattering* or *anomalous dispersion*. Because of its phase shift, this additional scattering may be divided into a real part $\Delta f'$ and an imaginary part $\Delta f''$. The real part can have either a positive or a negative sign; the imaginary part is always positive, i. e. the anomalous part always *advances* the phase angle somewhat. The result of this can more readily be visualized by considering the summing of the atom scattering factors in the complex plane (cf. Fig. 5.7). The normal scattering factor vector for the atom defines the origin for the anomalous vector. The real component simply lengthens or shortens the vector. The imaginary component is perpendicular to it, and *always* in the anticlockwise sense.

Since the "normal" scattering and the real part of the anomalous scattering are always parallel, f and $\Delta f'$ can always be combined for use in structure factor calculations:

$$f' = f + \Delta f' \tag{10.1}$$

The size of the anomalous effect is shown in Table 10.1. In particular, note the influence of the absorption edges for Co (which is just above the wavelength of Cu K_α radiation) and Ni (which is just below it). Since the absorption and anomalous scattering effects are closely related, a high anomalous scattering also implies a high atomic absorption coefficient (see sect. 7.4.3).

It is important that anomalous effects be taken into consideration, at least for atoms heavier than C with Cu radiation or heavier than Na for Mo radiation. In contrast to normal scattering, the magnitude of the anomalous scattering is *independent of the scattering angle*. This means that the effect is more pronounced at high values of ϑ. The actual measurable effects of anomalous scattering on the measured intensities depend on the space group of the crystal.

Centrosymmetric space groups. Without anomalous dispersion, the imaginary parts of the scattering vectors derived from a pair of atoms related by an inversion center are equal and opposite, and the phase angle Φ of the resultant must be either 0 or 180° (Fig. 10.5a and sect. 6.5.4) Since the imaginary parts of any anomalous scattering have the same sign, however, there must be a resultant phase angle, despite the presence of the inversion center. The amplitudes of the contribution of the two atoms remain equal. The Friedel Law, $|F_c(hkl)| = |F_c(\bar{h}\bar{k}\bar{l})|$, does still apply. This is easy to show by consideration of the structure factor equation:

$$F_c = \sum f_i \left\{\cos\left[2\pi(hx_i + ky_i + lz_i)\right] + i \sin\left[2\pi(hx_i + ky_i + lz_i)\right]\right\}$$

An atom (1) at xyz will make the same contribution to the reflection $|F_c(hkl)|$ that the atom (2) at $\bar{x}\bar{y}\bar{z}$ makes to the reflection $|F_c(\bar{h}\bar{k}\bar{l})|$, since the inversion of all the signs will leave the products unchanged

$$\begin{aligned} F_{hkl}(1) &= F_{\bar{h}\bar{k}\bar{l}}(2) \\ \text{and}\quad F_{\bar{h}\bar{k}\bar{l}}(1) &= F_{hkl}(2) \end{aligned} \tag{10.2}$$

(For the single atom i, $F_{hkl}(i) = f_i(\cos\Phi_i + i\sin\Phi_i)$.)

Table 10.1. Anomalous dispersion factors $\Delta f'$ and $\Delta f''$ and mass absorption coefficients μ/ϱ for Cu- and Mo-K_α radiation and a selection of common types of atoms. From *Int. Tab. C* [12] Tables 4.2.6.8 and 4.2.4.3.

	CuK_α			MoK_α		
	$\Delta f'$	$\Delta f''$	μ/ϱ [cm²/g]	$\Delta f'$	$\Delta f''$	μ/ϱ [cm²/g]
C	0.0181	0.0091	4.51	0.0033	0.0016	0.576
N	0.0311	0.0180	7.44	0.0061	0.0033	0.845
O	0.0492	0.0322	11.5	0.0106	0.0060	1.22
F	0.0727	0.0534	15.8	0.0171	0.0103	1.63
Na	0.1353	0.1239	29.7	0.0362	0.0249	3.03
Si	0.2541	0.3302	63.7	0.0817	0.0704	6.64
P	0.2955	0.4335	75.5	0.1023	0.0942	7.97
S	0.3331	0.5567	93.3	0.1246	0.1234	9.99
Cl	0.3639	0.7018	106.	0.1484	0.1585	11.5
Cr	−0.1635	2.4439	247.	0.3209	0.6236	29.9
Mn	−0.5299	2.8052	270.	0.3368	0.7283	33.1
Fe	−1.1336	3.1974	302.	0.3463	0.8444	37.6
Co	−2.3653	3.6143	321.	0.3494	0.9721	41.0
Ni	−3.0029	0.5091	48.8	0.3393	1.1124	46.9
Cu	−1.9646	0.5888	51.8	0.3201	1.2651	49.1
As	−0.9300	1.0051	74.7	0.0499	2.0058	66.1
Br	−0.6763	1.2805	89.0	−0.2901	2.4595	75.6
Mo	−0.0483	2.7339	154.	−1.6832	0.6857	18.8
Sn	0.0259	5.4591	247.	−0.6537	1.4246	31.0
Sb	−0.0562	5.8946	259.	−0.5866	1.5461	32.7
I	−0.3257	6.8362	288.	−0.4742	1.8119	36.7
W	−5.4734	5.5774	168.	−0.8490	6.8722	93.8
Pt	−4.5932	6.9264	188.	−1.7033	8.3905	107.
Bi	−4.0111	8.9310	244.	−4.1077	10.2566	126.

This relationship still holds in the presence of anomalous scatterers, since the two atoms make an identical contribution even though they cause a phase shift $\Delta\Phi$. Calculation of the structure factor for the Friedel pair $F(hkl)$ and $F(\bar{h}\bar{k}\bar{l})$ for this two atom structure than gives:

$$
\begin{aligned}
F_c(hkl) &= F_{hkl}(1) + F_{hkl}(2) \\
F_c(\bar{h}\bar{k}\bar{l}) &= F_{\bar{h}\bar{k}\bar{l}}(1) + F_{\bar{h}\bar{k}\bar{l}}(2)
\end{aligned}
\qquad (10.3)
$$

Since the order of operations does not effect the result of vectorial addition, it is thus clear that the Friedel Law

$$
|F_c(hkl)| = |F_c(\bar{h}\bar{k}\bar{l})|
$$

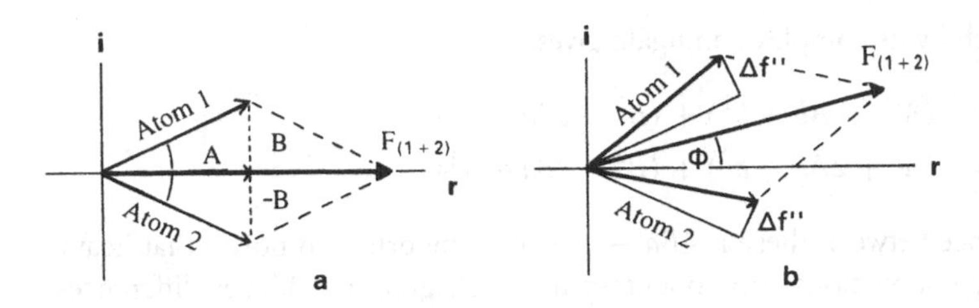

Fig. 10.5. Structure factors for a pair of atoms related by an inversion center (Atom 1 x, y, z; Atom 2 $\bar{x}, \bar{y}, \bar{z}$). **a** without **b** with anomalous dispersion.

is always valid in centrosymmetric space groups. Even so, it is essential in centrosymmetric structures to make the appropriate contributions to the structure factors of $\Delta f'$ and $\Delta f''$ for all heavy atoms. Most computer programs in use today do this automatically, at least for Cu and Mo radiation.

*Non-centrosymmetric space groups.*In the absence of an inversion center, the Friedel Law no longer applies when there is significant anomalous dispersion (Fig. 10.6). Inverting the indices now changes the sign of the "normal" imaginary contribution to the scattering, but not that of the anomalous contribution. Thus the vectorial additions lead both to different amplitudes and different phases for F_{hkl} and $F_{\bar{h}\bar{k}\bar{l}}$. The resulting differences in intensity between Friedel reflections, the *Bijvoet differences,* may be calculated in the following way.

Let the normal and anomalous contributions be divided, so that A and B represent the real and imaginary parts respectively arising from $f + \Delta f'(f')$, and a and b the anomalous real and imaginary parts, arising from the $\Delta f''$ terms for all atoms present. Since the contribution of $\Delta f''$ is always 90° in advance of the other, the structure factors may then be written:

$$F(hkl) = (A - b) + i(B + a)$$

$$F(\bar{h}\bar{k}\bar{l}) = (A + b) - i(B - a)$$

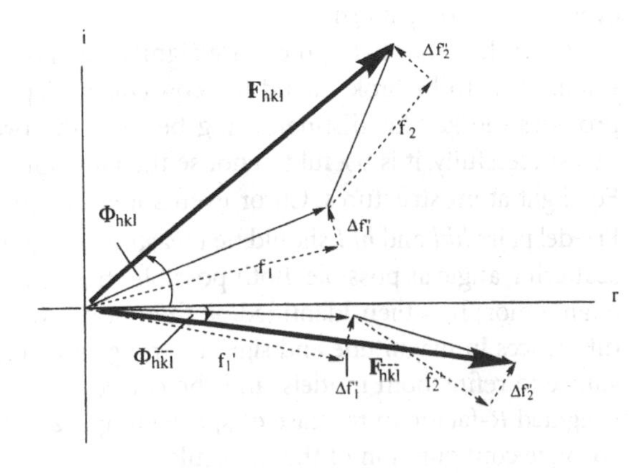

Fig. 10.6. The result of anomalous dispersion in a non-centrosymmetric space group.

multiplying each by its complex conjugate gives:

$$\begin{aligned} F^2(hkl) &= (A^2 - 2Ab + b^2) + (B^2 + 2Ba + a^2) \\ F^2(\bar{h}\bar{k}\bar{l}) &= (A^2 + 2Ab + b^2) + (B^2 - 2Ba + a^2) \end{aligned}$$

and the difference between these is $4Ba - 4Ab$. It is important to notice that heavy atoms with large anomalous components can cause significant Bijvoet differences even if they give a centrosymmetric arrangement themselves, provided that the *rest* of the structure deviates significantly from centrosymmetric. This situation is relatively common, as a structure in $P2_1$ with a single heavy atom in the asymmetric unit will always have a centrosymmetric arrangement of heavy atoms.

All of this implies that in non-centrosymmetric space groups, the Laue symmetry is not strictly observed and Friedel pairs I_{hkl} and $I_{\bar{h}\bar{k}\bar{l}}$ should not be averaged if there is any significant anomalous scattering. Note, however, that every space group (except $P1$) does have reflections which are genuinely related by symmetry and should be merged. A full table of these is given in *Int. Tab. B*, table 1.4.4.

The small intensity differences resulting from anomalous scattering are normally at best barely discernible in data from film or diffractometer. Only at the end of a structure determination, when the model is fully refined, can the anomalous scattering be properly assessed, but it is essential to consider it then. A simple example in space group $P2_1$ will be used to illustrate the effect.

Chiral, enantiomerically pure compounds can crystallize in space group $P2_1$, since it has neither an inversion center nor a mirror plane. It is, in fact, a very common space group for optically active natural products. In the absence of anomalous scattering, the diffraction pattern shows the symmetry of $P2/m$. If a structure has been refined as one enantiomer, say the R-form, the S-form will fit the data equally well, since inverting the structure at the origin will convert the xyz-parameters of all atoms to $\bar{x}\bar{y}\bar{z}$. In monoclinic crystals, the same effect can be achieved by reflection in the $\boldsymbol{a}, \boldsymbol{c}$-plane, i.e. by inversion of the y-parameter. These symmetry operations ($\bar{1}$ or $.m.$) are operations of the Laue group, and the diffraction is identical for the two enantiomeric structures. Put the other way around, X-rays cannot distinguish between them, the reason being the Friedel Law, which adds an inversion center to every diffraction pattern.

If, on the other hand, there are significant anomalous scatterers, this causes the Friedel Law to be broken, and the consequent departure from the Laue symmetry provides the key for distinguishing between the two enantiomers. In order to do this successfully, it is useful to choose the radiation in order to maximize the effect. For light atom structures, Cu or even softer radiation should be used. In addition, Friedel pairs hkl and $\bar{h}\bar{k}\bar{l}$ should be measured as carefully as possible and to as high a scattering angle as possible. Both possible structures are considered, and the correct enantiomorph is then identified as the one for which the data best fit the Bijvoet differences in magnitude and sign. If strong anomalous scatterers are present it may suffice to refine both models, and the correct one should give a significantly better weighted R-factor. In the case of space group $P2_1$, this will then identify the correct absolute configuration of the molecules.

10.2.1 Chiral and Polar Space Groups

Chiral molecular structures can only be described by those space groups which contain no symmetry elements involving inversion or reflection. As in the example above, a check has to be made for the "chirality" of the crystal. This is usually possible by simply inverting coordinates through the origin. The most important (commonest) space groups of this type are $P2_12_12_1$, $P2_1$, $P1$, $C2$ and $C222_1$ In cases involving 3_1-, 4_1-, 6_1- or 6_2-axes, the sense of screw must also be inverted, giving the enantiomorphic space group. Thus $P3_1$ becomes $P3_2$, $P4_1$ becomes $P4_3$ etc.

The other non-centrosymmetric space groups, which have mirror planes or $\bar{4}$- or $\bar{6}$-axes, e.g. $Pna2_1$, $Imm2$, or $I\bar{4}c2$, are often described generally as "polar" although the name properly describes only some of them as well as some of the chiral space groups. In them, the inversion of the direction of one axis (e.g. z in $Pna2_1$) leads to an arrangement of atoms distinguishable from the first by its effect on the anomalous scattering, even though neither group is chiral and any group of atoms that is present is accompanied by its mirror image. Since the initial choice of unit cell was arbitrary, it is essential at the end to examine the alternative orientation relative to the polar axes. Thus, in all non-centrosymmetric structures an "absolute" structure can in principle be determined which may be related to chirality, polarity or both. For a description of terminologies see [36, 37].

For pure organic compounds containing no atoms heavier than O, determination of absolute configuration is difficult even with the use of Cu-radiation. The differences in R-values are usually 0.001 or less. The presence of a single S-atom makes a large difference and the distinction is then usually made easily. In the presence of heavier atoms, the R-factor difference may be as much as 0.03, and it is possible to make the distinction even when Friedel pairs have not been measured.

The significance of the R-factor difference should be tested, using the weighted R values. Hamilton [38] proposed a test (in terms of the numbers of reflections and parameters of the two models) to estimate the probability e.g. 95 %, that one was actually correct. This mathematical model, of course, assumes that only statistical errors affecting a data set need to be considered. Today, when statistical errors are generally small with respect to systematic ones and data sets are very large relative to numbers of parameters, this test normally indicates that nearly any improvement in the R-factor is significant. It should be used with caution, and consideration made of any likely systematic errors in the data or the model.

In those cases where a clear distinction can be made, the refinement of the "wrong" absolute structure actually leads to systematic errors in atomic parameters and in the bonds and angles derived from them. The determination of the absolute structure is thus essential for obtaining an accurate molecular geometry, even when the absolute structure itself is of no interest, and other crystals in the same batch may have the alternative structure.

Inversion or racemic twinning. Sometimes, there is very little difference between the wR-values for the two alternative structure models even though strong anomalous scatterers are present. This can be explained in terms of *inversion twinning* (for a discussion of twinning in general, see section 11.2). There are regions – more properly domains – in the crystal of one absolute structure and domains of the other which have grown together in such a way that the crystallographic axes coincide, with one or more of them having opposite directions. A typical example occurs in monoclinic crystals, e.g. in space group $P2_1$, where twin domains may be related by a mirror plane normal to ***b*** (Fig. 10.7). Thus both enantiomorphous structures are present at the same time. In any non-centrosymmetric space group, the same effect can occur as a result of inversion of axial directions in the origin.

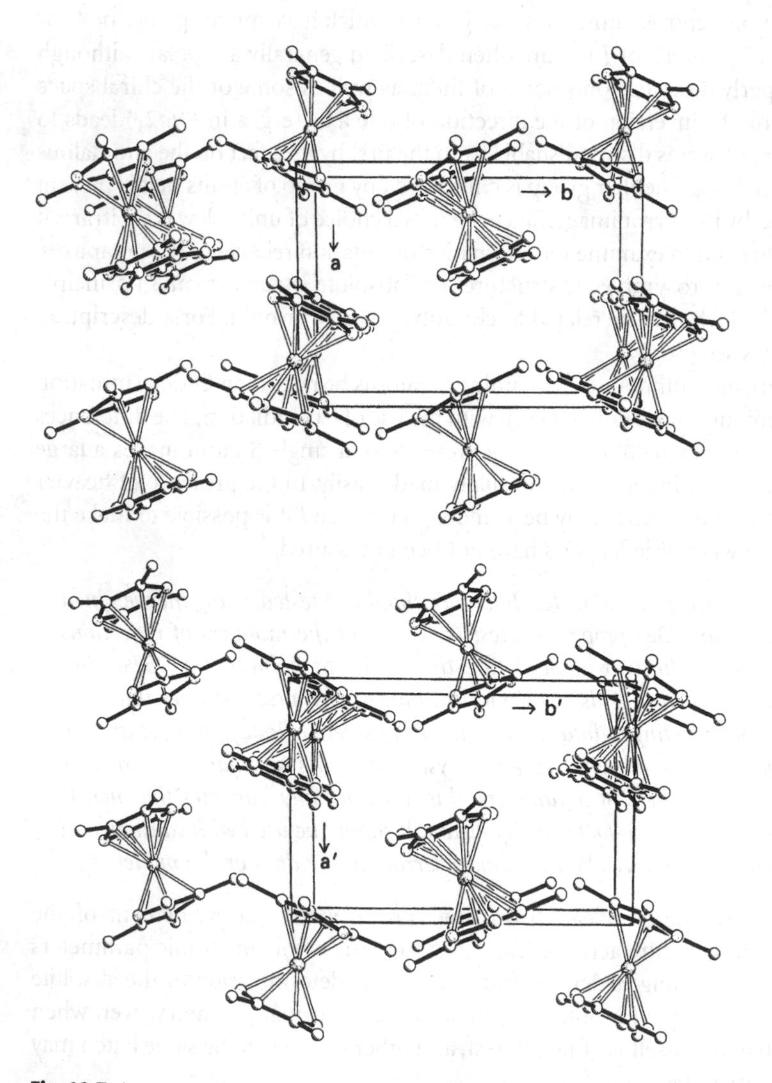

Fig. 10.7. A non-centrosymmetric structure (space group $P2_1$) (above) together with its mirror image (below) arranged to show a possible inversion twinning.

The diffraction pattern gives no obvious clue to the occurrence of inversion twinning, since a mirror plane normal to (010) as well as an inversion center are implicit in the Laue group $2/m$. In the absence of anomalous scatterers, the twinning is not detectable and does not affect the structure determination. It can only be detected in those cases where there is a large calculated anomalous effect but the determination of absolute structure is still impossible or very difficult.

Inversion twinning is, however, possible to quantify in terms of the structure model. An elegant method for that purpose was proposed by Flack [39]. It is accomplished by adding a parameter x to those being refined such that the calculated structure factor contains a fraction $1 - x$ of the model being refined, and x of its inverse. Provided that a low standard error is obtained, a value of 0 indicates that the model is correct, while a value of 1 that the inverse is correct and the structure has to be inverted before further refinement. If a value is somewhere in between, with a small error, that indicates the presence of inversion twinning, and a $1 : 1$ ratio is indicated by $x = 0.5$. In this case refinement of a twin model is obligatory. Disregarding inversion twinning may cause small errors in the structural geometry.

10.3 Extinction

A further important effect must be considered in the final stages of refinement which reveals its presence when reflections of high intensity and low scattering angle systematically give $|F_o| < |F_c|$. This is usually due to *extinction,* which can be divided into *primary* and *secondary* extinction. They are most significant for crystals of high quality. As has already been mentioned in section 7.2.3, real crystals have a mosaic structure that causes the diffracted beam to have a higher divergence and a lower coherence than the incident beam, so that it leaves the crystal without being further diffracted itself. A real crystal with significant and uniform mosaic spread is said to be *ideally imperfect,* and the *kinematic* scattering theory previously described applies to it. In such cases, the reflection intensities

$$I_{hkl} \sim F_{hkl}^2. \tag{10.4}$$

the more nearly a crystal approximates to ideality *without* mosaic spread, the more intense is the diffracted beam, and the more likely it is to function as the "primary beam" for further scattering.

Primary extinction occurs when a strong diffracted ray does function significantly as a primary beam (Fig. 10.8) and is thus weakened by further diffraction. As such a crystal approaches ideality, the intensity of the reflection is reduced until

$$I_{hkl} \sim | F_{hkl} | \tag{10.5}$$

the theoretical description of these scattering phenomena require another approach, that of *dynamic* scattering theory. Such near ideality is very rare in real crystals and occurs mainly in extremely pure materials such as semiconducting Si and Ge. For

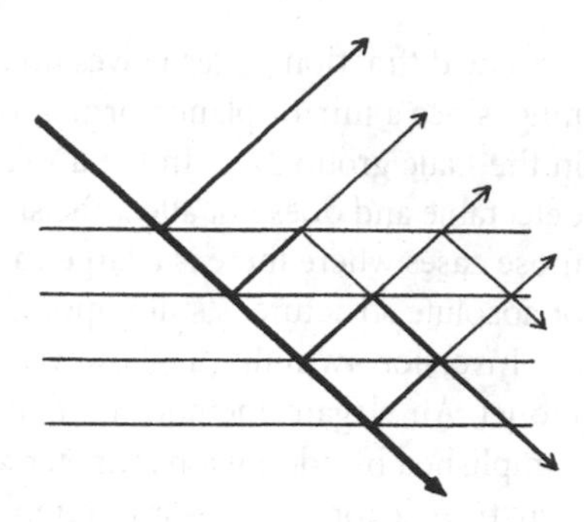

Fig. 10.8. Primary extinction.

crystals originating in a chemical laboratory, the kinematic theory is quite sufficient, and any small amount of primary extinction can be corrected by an empirical factor.

Secondary extinction occurs when, as the primary beam proceeds through the crystal, a significant fraction is diffracted by the first layers encountered (Fig. 10.9). This means that planes further into the crystal experience a reduced primary beam, and the overall effect is that the reflection is weakened. In an "ideally imperfect" crystal, the intensity of the primary beam is little affected by the scattering (less than 1 %) and the effect can be ignored. Secondary extinction is, however, much commoner that primary, and various theories have been devised to deal with it [40, 41], review [42]. Since the two effects are difficult to separate, in "normal" structure analyses, an empirical correction factor ε is applied to the F_c-values, and is refined with the other parameters. In the SHELXL program, this correction takes the form:

$$F_c(\text{corr}) = \frac{F_c}{(1 + \varepsilon F_c^2 \lambda^3 / \sin 2\vartheta)^{1/4}} \qquad (10.6)$$

where, as usual, λ is the wavelength of the X-rays, and ϑ is the scattering angle.

Extinction is more important with Cu-radiation than with Mo, since the proportion of the radiation scattered is greater. If the extinction is very large (differences between F_o and F_c greater than say 20 %) it can sometimes be decreased by dipping the crystal quickly into liquid nitrogen, when the thermal shock will often increase the mosaicity of the crystal.

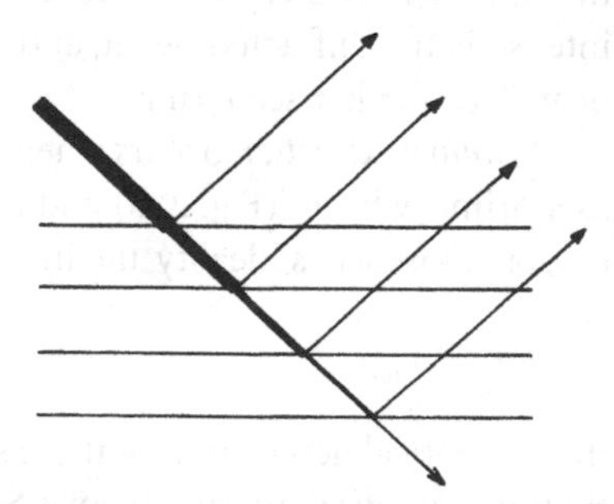

Fig. 10.9. Secondary extinction.

10.4 The Renninger Effect

An effect which is often ignored is the *double diffraction* phenomenon, normally called the *Renninger effect* after its discoverer [43]. It occurs in the following way. When a set of planes (hkl) with scattering angle ϑ is in the appropriate position to diffract the X-ray beam into the counter positioned with the correct value of 2ϑ, there is a reasonable probability that there is (at least) a second set of planes (with a different ϑ-value) that is also in the diffracting position. Since, however, X-rays diffracted from such planes will not enter the counter, they are usually of little significance.

It is, however, possible, that a third set of planes will be in the correct position to scatter radiation from the second set, and further that the direction of this twice diffracted beam is in the right direction to enter the counter. The reason for this is the indices of the first two sets of planes sum to those of the third, in a manner analogous to the triplet relationship in direct methods (Fig. 10.10).

The probability that the geometric conditions exist for such double diffraction is surprisingly high. There are normally for each reflection a considerable number of possible alternative paths. The results are only important, however, when both of the planes taking part in the double diffraction diffract very strongly and the crystal is of high quality, such that the doubly diffracted beam is actually strong enough to be measured. Only a few reflections will meet these conditions. Consequently, the effect will cause only an insignificant error if the reflection actually being measured is itself strong. On the other hand, it is quite possible that the "original" reflection is a systematic absence. In this case, the Renninger-reflection is observable, does not obey the diffraction conditions, and can lead to a wrong choice of space group.

Renninger reflections can usually be identified by the fact that the double diffraction reduces their half-width and makes them much sharper than ordinary reflections. On a diffractometer, they can be identified by doing a ψ-rotation about the normal to the "original" lattice planes. If the reflection is "genuine" its intensity will remain nearly constant. If, on the other hand, it is a Renninger reflection, it will vanish after a rotation of 1–2 , since the geometric conditions of Fig. 10.10 will no longer

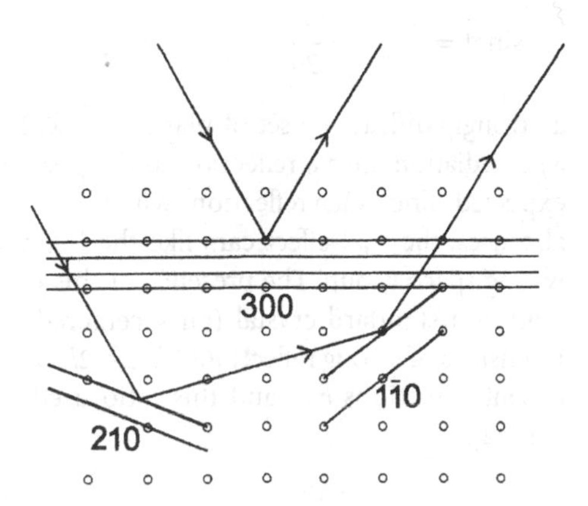

Fig. 10.10. The Renninger Effect.

hold. If it is desired to exclude completely the interference of Renninger effects, each reflection can be measured at two azimuthal angles (say 0° and 5°) in order to check that the intensities are the same.

The Renninger effect should always be considered when a structure could be described in a space group of higher symmetry which has been excluded by the presence of a few weak exceptions to the required systematic absences. Cases have been reported in which over 40 Renninger reflections were observed in positions corresponding to systematic absences. The lengthy controversy over the "correct" space group for the orthorhombic mineral weberite (*Imma*, *Imm*2, or $I2_12_12_1$) was eventually settled by the demonstration that all the exceptions to the systematic absences required for the *a*-glide in *Imma* could be assigned either to the Renninger effect [44, 44a] or to the $\lambda/2$-effect described below.

10.5 The λ/2-Effect

Large, strongly diffracting crystals can show a strong and troublesome "$\lambda/2$-effect". This results from the fact that the monochromatization of the X-rays may result in including a small amount of radiation with a wavelength half of that being sought. This can easily happen at a graphite monochromator, since, in terms of the Bragg-equation

$$2d \sin \vartheta = n\lambda$$

the nth order of diffraction (e.g. $n = 1$) for the desired K_α-radiation will appear at the same diffraction angle ϑ as the $2n^{\text{th}}$ order of diffraction (e.g. $n = 2$) for radiation with *half* of that wavelength, $\lambda/2$.

Whether there is any significant effect from radiation with such a wavelength depends on the type, quality and installation of the counter. With graphite monochromatization, generally, the quantity of $\lambda/2$-radiation is about 0.1–0.3 % of the total (for CCD detectors see [45]). Because of the equivalence:

$$\sin \vartheta = \frac{\lambda/2}{d} = \frac{\lambda}{2d}$$

a strongly diffracting set of planes $2h, 2k, 2l$ in the crystal (say 200) can then, with $\lambda/2$-radiation, give a reflection in the position where the first order hkl (say 100) is expected. Since such reflections with odd indices frequently correspond to systematic absences, the $\lambda/2$-effect can, like the Renninger effect lead to the assignment of a wrong space group. The presence of this effect is, however, easy to correct with the help of a standard crystal (an I-centered one is particularly useful). Using it, the intensities of strong reflections $2h, 2k, 2l$, can be compared with those of theoretically absent reflections hkl, and this ratio used to subtract out the $\lambda/2$-contribution in general.

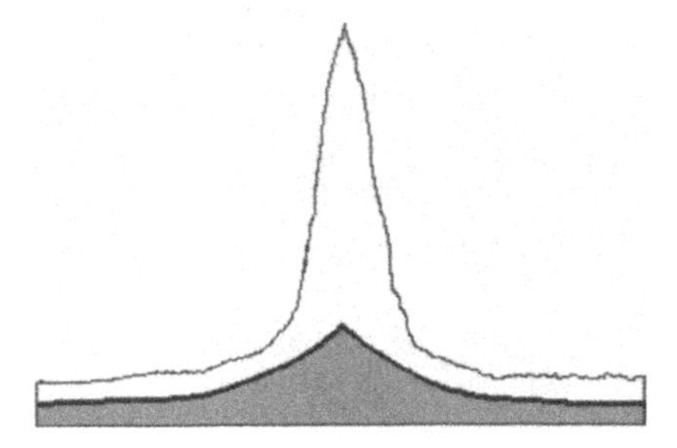

Fig. 10.11. The contribution of thermal diffuse scattering to a reflection.

10.6 Thermal Diffuse Scattering (TDS)

In Section 5.2, it is shown that the scattering of X-rays is strongly affected by the thermal vibrations of the atoms. The contribution of the vibrations to the form factor f is allowed for by the use of the displacement parameter U. There is, in fact, a further contribution to the scattering resulting from long-range vibrations, the lattice vibrations of the crystal. These lead to a variable diffuse background over the entire reciprocal lattice, and give rise to broad peaks with sharp maxima in the region of the Bragg reflections (Fig. 10.11). Because of these peaks, the subtraction of background from a peak intensity as described in section 7.4 does not correct fully for this effect. In fact, these peaks can, at high scattering angle, contribute up to 25 % of the net intensity. Since, however, these errors only have an effect on the displacement parameters, and rarely affect the refined atomic positions, TDS corrections are rarely made. When accurate electron density is required, a correction for TDS must be made, and methods for doing this are described in [46].

CHAPTER 11

Errors and Pitfalls

When a crystal structure has been successfully solved and refined by the methods described here, the resulting information is generally of a high level of accuracy which can only very exceptionally be obtained by spectroscopic methods. The indirect nature of the structure determination, depending as it does on the acceptance of a structure *model*, can, however, sometimes lead to serious errors. Some very nasty traps lie in wait for those who never look carefully at photographic or area detector data, and whose unit cells and space groups are determined by automatic diffractometers and are not critically considered! It is, of course virtually unthinkable that a completely nonsensical structure model could actually yield calculated F_c-values for several thousand reflections which agreed with the observed F_0-values within a few percent. On the other hand, there certainly are situations where the structure model largely reflects reality, and gives a credible R-factor although some small, but possibly important, parts of the structure are wrong. These can arise, for example, when the refinement converges not to the *global minimum* but to a local *false minimum*. In such a case, a solution has been obtained which contains errors which prevent the true structure being attained by further refinement in the normal way. There are other reasons for this to occur than a wrong choice of unit cell of space group, and specific cases of these will be examined in the following sections.

11.1 Wrong Atom-Types

The scattering power of an atom for X-rays is, as was explained in chapter 5, proportional to its atomic number. This means that neighboring atoms in the periodic table differ only slightly from one another. This is generally not a problem, since a chemist can generally identify an atom from its geometry – from the length, number, and arrangement of the bonds it makes. For good data sets, it is usually possible to make this distinction from the X-ray data alone, as the example pictured in Fig. 11.1 shows. In this case, a nitrogen atom in a complex was accidentally refined as carbon. Even though this error is only an error of one electron in the 171 in the structure model, correcting the atom type led to a significantly lower R-factor and a significantly more sensible atom displacement parameter.

It is much more difficult to distinguish such alternative possibilities when the data set is bad, since the differences both in the R-factors and in the bond lengths will be

atom "N1"	= N	= C
wR_2 (all refl.)	0.0753	0.0875
$R(I > 2\sigma)$	0.0297	0.0340
GOF	1.048	1.218
U_{eq} ("N1")	0.0320(7)	0.0203(8)
U_{eq} (C2)	0.0292(7)	0.0300(9)

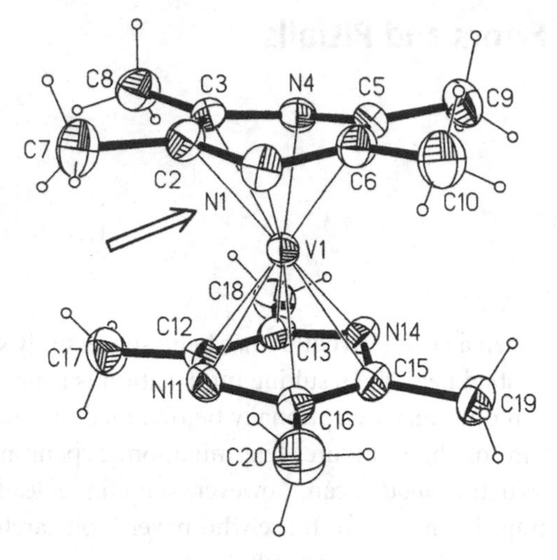

Fig. 11.1. Result of a wrongly-identified atom: "C" in place of N1 (arrow) in the sandwich complex Bis(tetramethyl-η^6-pyrazine)vanadium.

lost in the higher inaccuracy of the structure, and both chemical and crystallographic criteria will fail.

Sometimes an entirely unexpected product is synthesised, and there is a danger that the desire to find something really exciting will lead to an incorrect structure model to be chosen and normal healthy scepticism to be suspended! A particularly striking example of this type, where "the wish is father to the thought" occurred some years ago and led to the publication of a chemically completely wrong structure. This was corrected by von Schnering and Vu [47] who showed that a structure reported as the remarkable "$[ClF_6]^+[CuF_4]^-$" ($R = 0.07$) was actually the much less startling hydrolysis product $[Cu(H_2O)_4]^{2+}[SiF_6]^{2-}$. The substitution of Cl for Si and F for O was, in fact, detectable by the fact the displacement parameters for those atoms were 2–3 times the expected values. The clues to the discovery of this fault were the remarkable magnetic properties and blue colour of the salt. The lesson of this story is that whenever an unusual structure appears to have occurred, it is wise to check carefully the plausibility both of the displacement parameters of the atoms and of the chemical and physical properties of the crystals.

11.2 Twinning

Twinned crystals of minerals are very well known, such as the "swallowtail twin" of gypsum or the "Carlsbad twin" of orthoclase. When twinning occurs in crystals grown for a structure determination, however, it has usually been a good reason for proceeding no further with the work in view of the difficulties expected. On the other hand, unrecognized twinning has been an important cause of wrong structures being published. For this reason, the main types of twinning and their recognition and treatment will be briefly discussed here (see also E. Koch in *Int. Tab. C*, Section 3.1).

Twinning implies the growth of two or more differently orientated domains of a single structure into a twinned crystal according to a *twin law*. The two types of domain have a real axis (and consequently a reciprocal plane!) in common, or alternatively a real plane (and a reciprocal axis). The twinning can be described in terms of a symmetry element, the *twin-element*, which, unlike normal symmetry elements, does not occur in every unit cell but relatively few times — or even only once — on a macroscopic scale. The twinned crystal can be described and analysed from various points of view.

11.2.1 Classification by the Twin-Element

The twin-element is important, as it is the *twin-operation* which converts one part of the twin into the other.

Rotation twins. In this case, the common axis can be a 2-, 3-, 4- or 6-fold axis, and is usually, but not always one of the three axes of the unit cell (e.g. Fig. 11.2b).

Reflection twins. The commonest twins are of this sort, where the two domains are related to one another by reflection in a plane (hkl) (e.g. Fig. 11.2a) and are described as a "reflection twin in (110)," etc.

Inversion or racemic twins. Twins, in which a non-centrosymmetric structure and its mirror image have grown together are relatively common, but easy to handle. Since the Friedel law (sect. 6.5.4), whereby every diffraction pattern is centrosymmetric is very nearly obeyed by all structures, the reciprocal lattices of the two domains are virtually identical, and the structure may be solved and refined as either of the two mirror images. The existence of the twinning can only be detected by the application of anomalous scattering (sect. 10.2). This phenomenon can occur for many structures, and should always be investigated when an "absolute" structure has been determined.

11.2.2 Classification According to Macroscopic Appearance

Twinned crystals of minerals were recognized long before the development of X-ray analysis, and were classified according to the way they grew. For example, *contact twins*, where two separate crystals had grown outwards from the common twin plane can be distinguished from *interpenetrant twins* in which the two components appear to grow through one another (Fig. 11.2b). *Polysynthetic* or *lamellar* twins (Fig. 11.2c) grow as stacks of layers of the two components. If these are very thin (100–1000 Å) the existence of such *microscopic twinning* cannot be detected optically.

If the domains are big enough, and the structure is not cubic, the twinning can in principle at least be detected using a polarizing microscope (taking care not to look along the principal axis of a uniaxial crystal (cf. Section 7.1)). If a twin is detected, it is best first to try to find an untwinned specimen. Sometimes, an untwinned piece may be separated from the rest of a crystal using a scalpel, first immersing the crystal in an inert oil in order to prevent the desired fragment springing away! If all of the crystals are twinned and inseparable, the twin law should, if possible be determined, probably

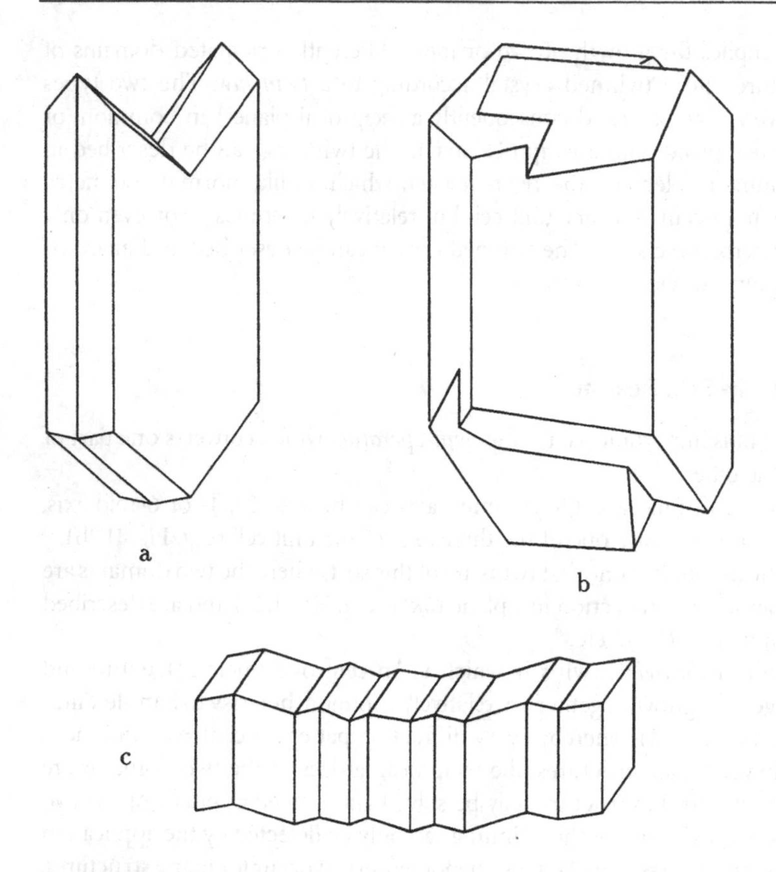

Fig. 11.2. Examples of twinned crystals: **a)** contact twin. **b)** interpenetrant twin and **c)** polysynthetic twin (effect greatly exaggerated).

by film or area detector methods. In many cases, it is quite possible to measure a data set and solve and refine the structure successfully.

"Multiple-twinning". This term is loosely applied to crystals with more than two types of twin-domain, which are more properly called *trillings, fourlings,* etc.

11.2.3
Classification According to Origin

Twins with large domains develop frequently from solutions or melts when crystals grow outwards from a single nucleus in more than one direction. These are called *growth twins.* In contrast, those microscopic, polysynthetic twins which develop when a crystal is grown at a high temperature and then undergoes a phase change on cooling from a higher to a lower symmetry are called *transformation twins.*

Twinning accompanying a phase change. If the space groups of the high- and low-temperature phases are known, the Landau theory of group/sub-group relations between the two space groups can be applied to predict the type of twinning that will occur. There are two types of symmetry loss on phase changes, known as t- and k-groups (see Sect. 6.4.5). An example of a change to a t-group is the degradation from *P*4/*nmm* to its subgroup *P*4/*n* with the loss of the (100), (010) and (110) mirror planes. Since the symmetry relaxation can occur with equal probability along the ***a***- or the ***b***-direction, twinning can occur in which a lost symmetry element (here (110)) becomes the *twin element*. Such twinning is inevitable, except when the crystal has been specially grown in an anisotropic environment, such as an electric field gradient, or below the temperature of the phase change, possibly by hydrothermal techniques. Clearly it is also possible to produce growth twins under these conditions.

In a change to a k-group, translational symmetry is lost, e.g. centering, but the crystal class is not changed. This can give rise to so-called *antiphase domains* which correspond to harmless inversion twins.

11.2.4 Diffraction Patterns of Twinned Crystals and their Interpretation

In the diffraction pattern, i.e. the weighted reciprocal lattice, of a twinned crystal, the reciprocal lattices of both domains overlap one another. The reciprocal equivalent of the real twin axis or plane can be directly applied to the reciprocal lattice of one component to generate that of the other. In the real lattice, the twin axis or plane may not pass through the origin of the unit cell; it may, for example lie normal or parallel to a plane of closest packed atoms. Since, however, the reciprocal lattice is independent of the choice of origin, the reciprocal equivalent of the symmetry element may be placed at the origin of the reciprocal lattice. Three types of twinning can be distinguished:

Non-merohedral twinning. (Twinning without coincidence in the reciprocal lattice). In these cases, there is no complete overlapping of the reflections for the two components, which is what is implied by *merohedry*. The twin element is *neither* an element of the crystal class *nor* of the crystal system. Since the two sets of reflections are more or less separate, it is usually relatively easy to discover the twin law. It may then be possible to measure only those reflections corresponding to a single component, eliminating or correcting those to which the other makes a contribution, and to solve and refine the structure in the usual way. Most twinned crystals are of this type; Fig. 11.3 shows as an example a schematic diagram of the $h0l$-layer of the reciprocal lattice of a monoclinic twin about (100) with equal components.

Such twinning is easy to recognise on film or area detector. On serial diffractometers, indexing usually fails at first, as both components will normally contribute to the set of trial reflections. If it is possible to isolate a set for one component only, perhaps using a program like DIRAX [77], indexing can succeed. Once again, the value of a film or an area detector, looking at a section of the lattice and not a single point is very great, and helps to prevent the choice of a pseudocell. As in the example shown in Fig. 10.3, both components usually coincide on a zero layer of the reciprocal lattice, here $0kl$, which contains contributions from both. If the Laue symmetry is, as here

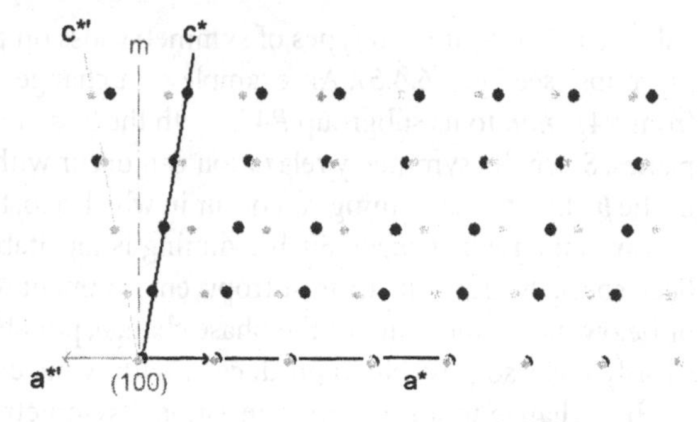

Fig. 11.3. Superposition of the *h0l*-layers of the reciprocal lattices in a non-merohedral monoclinic (100)-twin.

$2/m$ or higher, only symmetry equivalent reflections will coincide. If the intensity of a few strong data are known for both components of the twin, it is possible to derive a factor to scale these data to the rest of the data set and proceed normally with the structure determination.

Partial merohedral twinning. This type of twinning arises when, by chance, the unit cell parameters are such that in every second, third, or n^{th} layer the reciprocal lattices of the two components overlap. This is called *partial merohedry* with an index of 2, 3, or n.

This is a hazardous situation! If the presence of twinning is not known, as when all information comes from a serial diffractometer, it is possible to interpret the overlapping reciprocal lattices in terms of a smaller reciprocal cell, i.e. a larger real cell. The structure derived from this is wrong, but its errors may not be apparent. The usual hint that something is amiss comes from the occurrence of "impossible" systematic absences among the data. In the example shown in Fig. 11.4, indexed on the large cell, the condition for reflection hkl, $l = 2n : h = 2n$ would occur.

If the overlapping reflections are required for the structure determination, the relative contributions of the first component to the diffraction pattern, i.e. its contribution to the volume of the entire crystal, must be estimated: $x = V_1/(V_1 + V_2)$. This can often be done by using reflections which do not overlap. If the contributions of

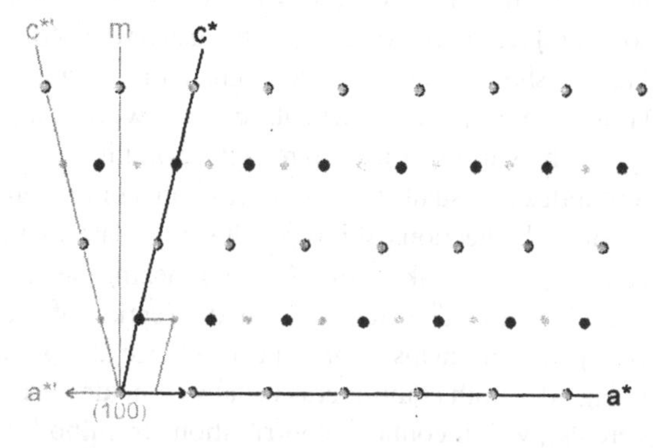

Fig. 11.4. Coincidence in every second layer along c^* in a partially merohedral twin. The apparent, too small reciprocal cell is outlined.

the two components are significantly different form 0.5, a mathematical separation of each reflection is possible (see example in [50]):

$$I_{hkl} = I_1 \frac{x}{2x-1} - I_2 \frac{1-x}{2x-1} \tag{11.1}$$

(I_1 and I_2 are the observed intensities for the (overlapped) reflection *hkl* indexed on the basis of the first and of the second components respectively.) After this treatment, structure determination can proceed as normal.

Merohedral twinning. When the twin-element is not a symmetry element of the crystal class but *is* an element of the crystal system (the crystal class of the lattice), this means that different orientations of both the direct and reciprocal lattices will coincide exactly with one another (Fig. 11.5) This is called *merohedral twinning*, or more strictly, twinning through merohedry. Merohedral twins frequently occur when a phase transition takes a high temperature form of high symmetry to a low temperature form of a lower crystal class.

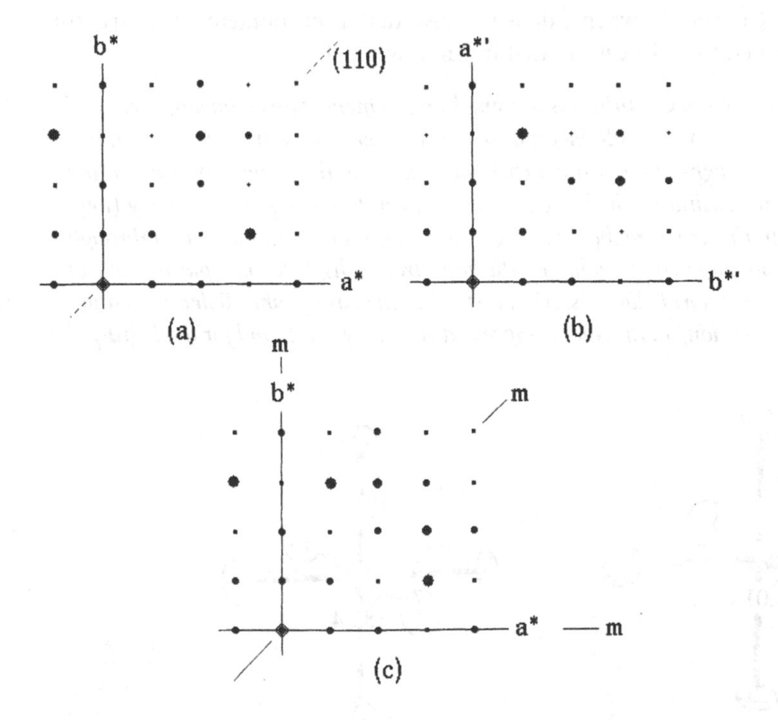

Fig. 11.5. Complete coincidence of the reciprocal lattices for a merohedral (110) twin of a tetragonal crystal with Laue symmetry $4/m$. **a)** $hk0$-layer of the first component with the twin element shown. **b)** The corresponding layer in the second component. **c)** Superposition of the two layers in a 1 : 1 ratio giving the Laue group $4/mmm$.

An example of this is the structure of $CsMnF_4$, in the tetragonal space group $P4/n$ [51]. The space group, and the crystal class $4/m$ have no symmetry elements along the **a**- *and* **b**-*directions or the* [110]-*diagonal. The tetragonal lattice, however, possesses the symmetry* $4/mmm$. *The crystal studied had a twin plane defined as* (110) — *or an equivalent arrangement using* (100). *These mirror planes are present in the lattice but not in the crystal class. They cause the lattice to be reflected into itself, such that all points in the reciprocal lattice fall on top of another. Thus, each reflection hkl from the first component falls on top of khl from the second. In the lower symmetry Laue group* $4/m$, *hkl and khl are not symmetry equivalent.*

If the contributions of the two components to the volume of the crystal are roughly equal, the weighted reciprocal lattice will appear to have the extra mirror planes and have the higher Laue symmetry $4/mmm$. This is, in fact, the true symmetry of the high temperature form. Neglecting the twinning will lead to a wrong space group of too high symmetry. It is sometimes possible, as in the last example, to solve and refine the structure in this space group, and obtain a false average of the two orientations of the actual structure. Frequently in such cases, disorder will be diagnosed. Conversely, when extensive "disorder" is encountered, the structure should be checked to make certain that it is not twinned.

As the example of $CsMnF_4$ has shown, the refinement of an erroneous "average" structure can lead to a good R-factor and a reasonable atomic arrangement (in this case that of the high-temperature form) when many of the heavy atoms lie in any case on special positions which themselves conform to the higher symmetry (here $P4/nmm$). In the structure shown in Fig. 11.6, the twinning is only apparent through the rather too large components of the anisotropic displacement parameters of the "equatorial" fluoride ligands. Also, instead of the strong Jahn-Teller elongation expected for a d^4 ion, an unusual compressed geometry was found for the $[MnF_6^{3-}]$

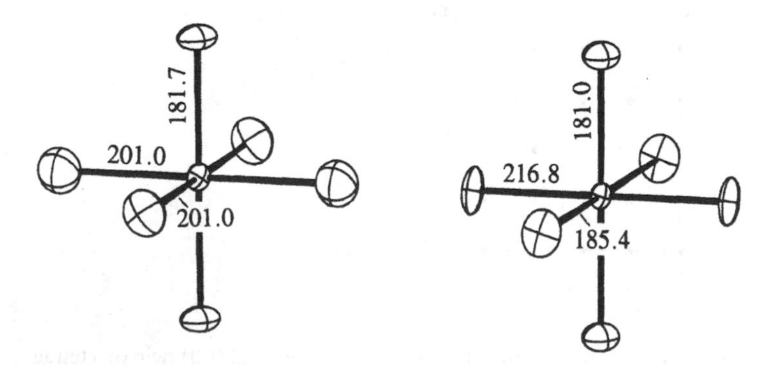

Fig. 11.6. *Example of distortion caused by an undiagnosed merohedral twinning ($CsMnF_4$):* ***a)*** *Details of structure after refinement in the apparent higher symmetry (space group $P4/nmm$, $wR = 4.85\,\%$).* ***b)*** *After refinement as a (110) twin in space group $P4/n$ ($wR = 4.39\,\%$) (Displacement ellipsoids have 50 % probability. Distances are given in pm (1 Å = 100 pm).*

octahedra, in disagreement with the ferromagnetic properties of the crystal. The refinement of the structure allowing for twinning led directly to the expected axial elongation of the octahedron, reasonable displacement parameters and a somewhat better R-factor.

If the contributions of the twin components of a merohedral twin differ significantly from 1 : 1, the weighted reciprocal lattice will show the *true* Laue symmetry and lead to the correct space group. Since the data set is, however, distorted by the presence of the minor component, the structure may not solve and even if it does, it will not usually refine well. In such cases, related structures must be used to provide a possible structure model, which can then be refined taking account of the twinning (see below).

In a few space groups, merohedral twins can be recognized from the fact that conditions for reflection occur which are not found in any space groups of the higher Laue group. Examples are $Pa\bar{3}$, $Ia\bar{3}$, $P4_2/n$ and $I4_1/a$.

Holohedral twinning. If the twin element is itself a symmetry element of the crystal class of the structure, the diffraction patterns of the two components are indistinguishable. Only symmetry equivalent reflections fall on top of one another. Such data sets behave entirely normally and need not be considered here.

Pseudomerohedral twinning. There are a few structures, whose lattice parameters almost meet the conditions for a higher symmetry, e.g. monoclinic crystals with $\beta = 90°$. In such a case, a twin plane on (100) will cause symmetry-unrelated pairs of reflections to lie too close to be distinguishable. If the two components are present in a 1 : 1 ratio, the weighted reciprocal lattice will show orthorhombic Laue group *mmm* instead of $2/m$, and the structure will not be solvable. If a model for the structure can be found, it can be refined as a twin (see below). If there is a slight splitting of the reflections, it is important that the scan is broad enough to encompass both components. In such cases, it may be necessary to adjust the orientation matrix to refer to the average position of the two components of the doublets. The determination of the unit cell, on the other hand, should make use only of data which can be assigned to a single component.

When the reciprocal lattices of two twin components are superimposed, it is often the case that a measurable reflection of one component will fall on top of a systematic absence of the other, and appear to contradict the conditions for diffraction. In such cases, if twinning is not suspected, a space group may be chosen with few translational symmetry elements, such as $C222_1$ or $Pmmm$. When such relatively rare space groups are encountered, particularly if the structure is difficult to solve, it is worth considering the possibility of twinning.

Refinement of twinned structures. The prerequisites for refining a *merohedral or pseudo-merohedral twinned structure* are the knowledge of the twin law and a suitable structure model for the untwinned structure. The "true" intensities for structure solution by Patterson or direct methods are not available because of the coincidence of the two reciprocal lattices. Two situations may be distinguished:

If the contribution of one twin component is so small that the twinning is barely observable, the correct space group will have been found and the structure solved normally. The twinning is only evident because the R-factor is higher than expected and there are small irregularities in the geometry of the structure and/or in the atomic displacement parameters. It is then necessary to determine the twin law and the matrix which will convert the axes of the first component into those of the second. In the tetragonal example above, this is:

$$\begin{pmatrix} 0 & 1 & 0 \\ 1 & 0 & 0 \\ 0 & 0 & -1 \end{pmatrix} \quad (11.2)$$

This matrix also serves to transform the indices hkl of the first component into the indices $h'k'l'$ of the second. Knowing the twin law, it is then possible to write the contribution of the second component to the structure factor F_c or F_c^2:

$$F_c^2(hkl)_{\text{twin}} = (1-x)F_c^2(hkl) + xF_c^2(h'k'l') \quad (11.3)$$

and then to refine that against the observed F_o^2(twin)-values. Refinement programs for twin- or trilling-structures are CRYSTALS [57] using F_o-data, SHELXL [68] using F_o^2-data, and JANA [79] using either.

If the twin components are nearly equal, an incorrect Laue group and space group will be indicated, and if the structure can be solved at all, it will be, as in the case of $CsMnF_4$ above be a false, averaged structure. If, however, the correct space group and a good model can be found, twin refinement as described above is possible. It is, however, necessary, if starting from a model of higher symmetry to "push" the structure off the false symmetry in one (often arbitrary) direction. In the above example of $CsMnF_4$, one pair of Mn—F bonds must be shortened and the other lengthened. A least squares program is normally unable to select one of two possible ways of lowering symmetry. In more complex structures, lowering the symmetry can involve a large number of possible combinations of atom shifts before one can be certain that the true, global minimum has been reached in the refinement, and not a local, pseudo-minimum. In such cases, it is particularly important to make use of other criteria than R-factors to judge the plausibility of the structure, especially the resulting geometry and the physical properties of the material.

Non merohedral or partially merohedral twins can be refined, using the HKLF5 option of the SHELXL program [68]. For this purpose, each reflection in the data set must be marked "1" or "2" to show to which twin domain it belongs. Overlapping reflections are entered twice, with the appropriate indices for each domain, but the total measured intensity for both. Partially overlapping reflections present a particular problem. If possible, the integration area on the detector system is made large enough to contain both components. Otherwise, these data are best omitted.

11.2.5 Twinning or Disorder?

Often, the evident anomalies of an "averaged" structure, such as high displacement factors and split atoms have been assigned to disorder (see above), even in cases where merohedral twinning is geometrically possible. In fact, the effects of the two phenomena are similar. In true disorder, there is a statistical distribution of the two alternative orientations over regions smaller than the coherence length of the X-rays, and the two contribute together to an average structure amplitude F_o. In a twin, the contributions of both twin domains are independent and, for overlapping reflections, the average of the individual F_o^2 values is observed. Which is the correct model can only be determined with certainty by trying both. When there is doubt, it is best to refine an ordered model, a disordered model and an ordered twin-model. The correct model should give not only a better R-value but also more reasonable atom displacement parameters. There are certainly many examples in the literature of merohedral twins which have been mistakenly refined as disordered structures.

Sometimes, doubts will arise about the correctness of what appeared to be a successfully solved and refined structure, but the crystal no longer exists, and a photographic examination cannot be made. In such cases, it is useful to make a transparent plot of the structure on an important plane of projection, and then to rotate or invert it on top of an identical diagram. In this way, possible twin planes or axes may be indicated. Whether they are, in fact, real can only be determined by painstaking comparative refinement of the alternative models.

11.3 False Unit Cells

As was mentioned above (section 7.2. and 7.3) one of the key steps in a structure analysis is the determination of the orientation matrix and the "correct" unit cell that is derived from it. The most serious source of error is failure to notice weak reflections, which can cause a doubling or tripling of one or more lattice parameters. This is particularly likely in the case when the crystals have superstructure properties (section 11.2.4) and a long photographic exposure or careful area detector investigation has not been made. Often, it is possible to measure data in this incorrect cell, to determine a (wrong) space group, and to solve and refine the structure. The result is an "averaged" structure in which, characteristically, alternate groups of atoms are superimposed on top of one another. The resulting effects, such as high displacement factors or splitting of atom positions are then wrongly ascribed to disorder.

On the other hand, various faults, including the $\lambda/2$ effect (Section 10.5) can cause a centered cell to appear primitive or one or more of the cell parameters to be doubled in length. This becomes clear in the refinement through the very high correlation between parameters which are actually related by a translation.

Another problem can arise simply because small deviations in lattice parameters encountered at the beginning of a structure determination are taken too seriously.

Errors in the centering of a crystal or of the goniometer itself can, for example, make one of the 90° angles in a monoclinic cell deviate a few tenths of a degree so that a triclinic cell is diagnosed. In such cases, the Laue symmetry must be checked before the crystal system is definitely fixed.

A similar misassignment occurs when a trigonal or hexagonal cell is originally described as an orthorhombic C-centered cell. It is always possible, using the matrix

$$\begin{pmatrix} 2 & 1 & 0 \\ 0 & 1 & 0 \\ 0 & 0 & 1 \end{pmatrix} \qquad (11.4)$$

to generate the "orthohexagonal" cell, which is apparently orthorhombic with $a/b = \sqrt{3}$.

If there is any suspicion that a false cell has been selected, recourse to photography or careful examination of area detector records is recommended! It is especially useful, if such a program is available, to examine the weighted reciprocal lattice layer by layer on the computer screen. The Laue symmetry and any possible systematic absences will then be visible, and can lead to a more sensible choice of axes and the merging of equivalent data.

11.4 Space Group Errors

A relatively common error, that often goes hand in hand with the problems described above is the choice of a wrong space group. There are two main cases that may be distinguished:

1. The unit cell is correct, but the wrong choice has been made among those groups which have the same systematic absences. In most cases, the choice is between a pair of space groups which differ in the presence or absence of a center of symmetry (Table 11.1).

Every centrosymmetric structure may, of course, be described in terms of a non-centrosymmetric group (e. g. in Cc instead of $C2/c$) by simply doubling the number of atoms in the asymmetric unit, and consequently the number of parameters to be refined. Since there are always errors in the measured data, the refinement in the space group of lower symmetry will always lead to a somewhat smaller R-value, but the large correlations between the parameters which are actually symmetry related will lead to high standard errors (cf. sect. 9.1). These correlations will lead to instability in the least squares refinement, and usually to obviously erroneous molecular geometry. In such cases, it is always important to test whether such an increase in the number of parameters really leads to a *significant* improvement. The most common cases of misassigned space groups of too low symmetry are given in Table 11.2.

2. If the unit cell is itself wrong, it is likely that the space group type will also be wrong. If, for example, a superstructure has been overlooked, a space group of too

Table 11.1. Selection of space groups which differ only in the presence or absence of an inversion center and hence have the same conditions for reflection.

triclinic	$P1$ (1)	$P\bar{1}$ (2)
monoclinic	$P2_1$ (4)	$P2_1/m$ (11)
	$C2$ (5)	$C2/m$ (12)
	Pc (7)	$P2/c$ (13)
	Cm (8)	$C2/m$ (12)
orthorhombic	$P222$ (16)	$Pmmm$ (47)
	$C222$ (21)	$Cmmm$ (65)
	$Pcc2$ (27)	$Pccm$ (49)
	$Pca2_1$ (29)	$Pcam$ ($\rightarrow Pbcm$) (57)
	$Pba2$ (32)	$Pbam$ (55)
	$Pna2_1$ (33)	$Pnam$ ($\rightarrow Pnma$) (62)
	$Cmc2_1$ (36)	$Cmcm$ (63)
	$Ama2$ (40)	$Amam$ ($\rightarrow Cmcm$) (63)
tetragonal	$I4$ (79)	$I4/m$ (87)
	$I422$ (97)	$I4/mmm$ (139)
	$I4mm$ (107)	$I4/mmm$ (139)
trigonal	$R3$ (146)	$R\bar{3}$ (148)
	$P3m1$ (156)	$P\bar{3}m1$ (162)
hexagonal	$P622$ (177)	$P6/mmm$ (191)

Table 11.2. The most common examples of wrongly assigned space groups in the literature (after [34]).

assigned space group	correct space group
Cc (9)	$C2/c$ (15), $Fdd2$ (43), $R\bar{3}c$ (167)
$P\bar{1}$ (2)	$C2/c$ (15)
$P1$ (1)	$P\bar{1}$ (2)
$Pna2_1$ (33)	$Pnam$($\rightarrow Pnma$) (62)
$C2/m$ (12)	$R\bar{3}m$ (166)
$C2/c$ (15)	$R\bar{3}c$ (167)
Pc (7)	$P2_1/c$ (14)
$C2$ (5)	$C2/c$ (15), $Fdd2$ (43), $R\bar{3}c$ (167)
$P2_1$ (4)	$P2_1/c$ (14)

high a symmetry is likely to have been assigned. Despite this, the structure can often be solved and refined to give an averaging of parts of the true structure.

If, because of errors in cell dimensions, the structure is solved in a crystal class and space group of too low symmetry, the refined structure model will contain two or more apparently independent molecules or other groups of atoms in the asymmetric unit. By studying their relation to one another, possibly using a program such as ADDSYM (former MISSYM) in PLATON [70], the missing symmetry elements can be found and the correct, more symmetrical space group established. According to *Baur* [34], the commonest errors of this sort occur when the true space group is $C2/c$, $R\bar{3}m$ or $R\bar{3}c$ (Table 11.2).

11.5 Misplaced Origins

Space group $P1$ or $P\bar{1}$? Most triclinic structures have an inversion center, and are thus in space group $P\bar{1}$. It is thus reasonable always to attempt first to solve a triclinic structure in $P\bar{1}$. There are, however, situations where difficulties in fixing the origin cause direct methods programs to fail to solve a structure, even though the correct space group, $P\bar{1}$ has been chosen. In such cases, a solution *is* often found in the corresponding non-centrosymmetric space group $P1$. In every structure solved in $P1$, it is wise to test whether an inversion center can be found somewhere in the refined structure model. In practice, perhaps based on a diagram of the structure, it may be demonstrable that the midpoint of all pairs of possibly symmetry related atoms lies at the same point xyz. In this case, subtraction of that xyz from the positional parameters of all atoms will shift the model to place the inversion center at the origin. Half of the atoms can then be removed from the list, and further work carried out in space group $P\bar{1}$.

Origin problems in space groups $C2/c$ and $C2/m$. In these relatively common space groups, there are alternative but non-equivalent sets of special positions the occupation of either of which gives the same array of atoms. In space group $C2/c$, one type are the inversion centers at sites $4a$ ($0, 0, 0; 0, 0, \frac{1}{2}; \frac{1}{2}, \frac{1}{2}, 0; \frac{1}{2}, \frac{1}{2}, \frac{1}{2}$) and $4b$ ($0, \frac{1}{2}, 0; 0, \frac{1}{2}, \frac{1}{2}; \frac{1}{2}, 0, 0; \frac{1}{2}, 0, \frac{1}{2}$) and the other are those at $4c$ ($\frac{1}{4}, \frac{1}{4}, 0; \frac{3}{4}, \frac{1}{4}, \frac{1}{2}; \frac{3}{4}, \frac{3}{4}, 0; \frac{1}{4}, \frac{3}{4}, \frac{1}{2}$) and $4d$ ($\frac{1}{4}, \frac{1}{4}, \frac{1}{2}; \frac{3}{4}, \frac{1}{4}, 0; \frac{3}{4}, \frac{3}{4}, \frac{1}{2}; \frac{1}{4}, \frac{3}{4}, 0$). If both sites $4a$ and $4b$ *or* both sites $4c$ and $4d$ are occupied, the same array of atoms will result, and the same R-factor will be calculated. Despite the fact that all of these sites have the same point symmetry, $\bar{1}$, the two sets are not crystallographically equivalent as they have different relationships to the rest of the symmetry elements of the space group. For example, sites $4a$ and $4b$ lie on an a-glide plane, while $4c$ and $4d$ lie on an n-glide. A similar trap occurs in space group $C2/m$, where the sites $2a-d$ with point symmetry $2/m$ taken together describe the same array as the sites $4e$ and $4f$ with $\bar{1}$ symmetry. Particularly in inorganic solids, heavy atoms often occupy such special positions, and at the beginning of a structure determination, it is possible to assign them to the wrong ones, so that the ligands surrounding them are not correctly described. This situation is usually indicated by the fact that the R-value drops to 0.2–0.3 and improves no further, and no fully complete and sensible structure develops in the Fourier syntheses. In such cases, shifting the origin by $\pm(\frac{1}{4}, \frac{1}{4}, 0)$ will shift the heavy atoms to the alternative positions and enable the structure to be completed.

11.6 Poor Atom Displacement Parameters

It has already been noted several times that physically unreasonable atom displacement parameters are often a sign that there is an error in the structure model. The shape of the displacement ellipsoids are indeed a very sensitive indicator for the correctness of a structure determination, often better than R-values or standard errors. What follows is a summary of the most probable causes of such poor displacement parameters.

- *Parallel orientation of displacement ellipsoids* with no obvious relationship to chemical structure usually indicates a poor or erroneous absorption correction for an anisotropic crystal form (e.g. a plate) and a high absorption coefficient (see Section 7.4.3).
- *A displacement parameter is very small or even negative.* This can indicate that a heavier atom should occupy this site (Section 11.1).
- *A displacement parameter is too large.* Either a lighter atom or no atom at all should occupy this site, or possibly due to disorder (Section 10.1), it is only partially occupied.
- *A displacement ellipsoid has a physically unreasonable form.* Displacement ellipsoids which are shaped like dishes or cigars or have no volume at all (i.e. are "non-positive definite") can indicate a poor data set (e.g. too low a data : parameter ratio), such that the anisotropic refinement of all of the atoms is not sensible. In such cases, isotropic displacement parameters should be used. The effect also often arises when an attempt is made to refine a disordered region with the help of a split-atom model. When the part-atoms of such a model come too close together, high correlations arise, and the displacement parameters are particularly sensitive to these. In such cases, it often helps to use common temperature factors for atoms in similar chemical environments or alternatively carefully chosen fixed values.
  The appearance of abnormal displacement parameters often indicates a more fundamental error in the structure determination, such as a wrong unit cell, a wrong space group or undiagnosed twinning (see above) which should be investigated further.

CHAPTER 12

Interpretation and Presentation of Results

Despite all the possible pitfalls and errors described in the last chapter, crystal structures usually arrive at the point where the refined model is ready to be converted into structural parameters of interest to chemists.

12.1 Bond Lengths and Bond Angles

The most important information is the separation of pairs of atoms, whether bond lengths or contacts. These are readily derived from the differences in the atomic coordinates:

$$\Delta x = x_2 - x_1 \quad \Delta y = y_2 - y_1 \quad \Delta z = z_2 - z_1$$

In order to convert these differences into absolute lengths, they must be multiplied by the lattice constants. The actual interatomic distance in the general, triclinic case is:

$$d = \sqrt{(\Delta x \cdot a)^2 + (\Delta y \cdot b)^2 + (\Delta z \cdot c)^2 + \cdots \atop \cdots - 2\Delta x \Delta y ab \cos\gamma - 2\Delta x \Delta z ac \cos\beta - 2\Delta y \Delta z bc \cos\alpha} \tag{12.1}$$

Bond angles are most readily calculated using the cosine law. The angle At2—At1—At3 in terms of the three interatomic distances d_{12}, d_{13} and d_{23} is:

$$\cos\varphi_{\mathrm{At2,At1,At3}} = \frac{d_{12}^2 + d_{13}^2 - d_{23}^2}{2d_{12}d_{13}} \tag{12.2}$$

The standard errors of the bond lengths, the most important error estimates in a structure, may be derived in a rather complicated way from the standard errors of the positional parameters of both atoms and the orientation of the interatomic vector, see [5, 10, 12]. A useful, if crude, approximation may be obtained by estimating the isotropic error for each atom ($\sigma_1 = (\sigma_x^2 a^2 + \sigma_y^2 b^2 + \sigma_z^2 c^2)^{1/2}$). Then, for a bond

$$\sigma_d = \sqrt{(\sigma_1^2 + \sigma_2^2)} \tag{12.3}$$

In most modern computer programs, the contribution of the standard errors of the lattice constants are also considered, although these are usually small enough to make an insignificant contribution to errors in bonds and angles.

An exception arises when both atoms lie on special positions with no free parameters, e.g. atom 1 at 0, 0, 0 and atom 2 at $\frac{1}{2}$, 0, 0: In such cases, the errors are due only to errors in the lattice constants.

In a well-determined light-atom structure, the standard errors for bonds between C, N and O-atoms are usually in the range 0.002–0.004 Å. For heavier atoms, these may be smaller than 0.001 Å.

In normal probability theory, the true value of a quantity should lie within ± 1.96 standard deviations of an estimated value with a probability of 95 %, and within ± 2.58 standard deviations with a probability of 99 %. This estimate, of course, considers only random errors and not systematic ones. A difference between two bond lengths should only be considered significant when it exceeds 3–4σ, therefore.

Corrections for vibrations. Groups of atoms which have large anisotropic motions, e.g. terminal —CO or —CN, sometimes give bond lengths that appear up to 0.05 Å shorter than expected. This arises from the fact that the oscillation of the group results in a banana-shaped electron density. Since vibrations are always described in the structure model as ellipsoids, the result is that the midpoint of the density is displaced along the bond toward the other atom (distance d' in Fig. 12.1). A simple theory has been devised to correct this by interpreting such an atom or group either as an oscillator (riding model) or as the vibration of a rigid group. In practice, such corrections are rarely carried out in published structures, but the effect should be kept in mind whenever "anomalous" short bond lengths are encountered.

12.2 Best Planes and Torsion Angles

Frequently, crystal structures are used to determine whether or not the coordination of an atom is planar or whether the conformation of a ring is flat or puckered. For this purpose, a least-squares method can be used to calculate a plane which minimizes

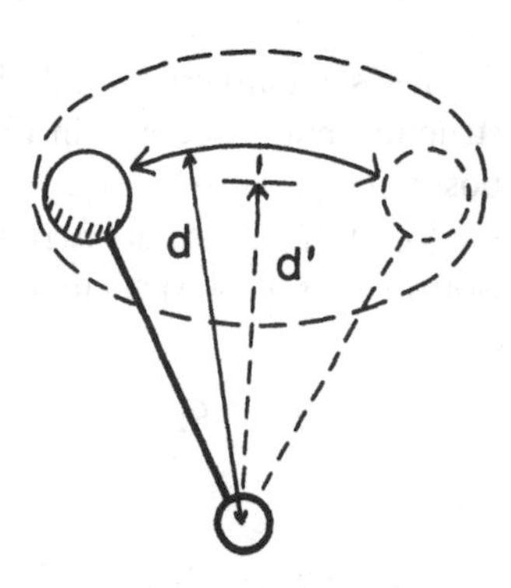

Fig. 12.1. Illustration of the apparent shortening of a bond length by strongly anisotropic vibration.

the sum of the squares of the distances δ of the i defining atoms from it (for full mathematical treatment, see *Int. Tab. B*, Sect. 3.2)

$$\sum_i \delta_i^2 = \text{min.} \tag{12.4}$$

The mean standard error of this "best" or "least-squares" plane is then

$$\sigma_\text{P} = \sqrt{\sum_i \frac{\delta_i^2}{i-3}} \tag{12.5}$$

and indicates how well the conformation of the group of atoms is described by a plane.

In discussing structures, *interplanar* or *dihedral* angles are often important. These are the angles between the normals to two planes. As an example, see the folding angles in the seven-membered ring depicted in Fig. 12.2, which considers the planes (1) defined by atoms 1, 2, 3 and 4, (2) by atoms 1, 4, 5 and 7, and (3) defined by 5, 6 and 7. The folding angles are then the supplements of the calculated dihedral angles.

A *torsion angle* is a special case of dihedral angle. As shown in Fig. 12.3, the torsion angle for the atoms 1–2–3–4 is essentially the angle between the planes defined by atoms 1, 2 and 3 and 2, 3 and 4. Alternatively, consider the angle between the bonds 1–2 and 3–4 projected onto the plane normal to the bond 2–3, with the bond 1–2 toward the viewer (a *Newman projection*). The torsion angle is then the angle measured from atom 1 to atom 4. If this is clockwise, the torsion angle has a positive sign, otherwise negative. This sign does not change if the measurement is made 4–3–2–1 instead, but the torsion angle for the mirror image of the group has the opposite sign.

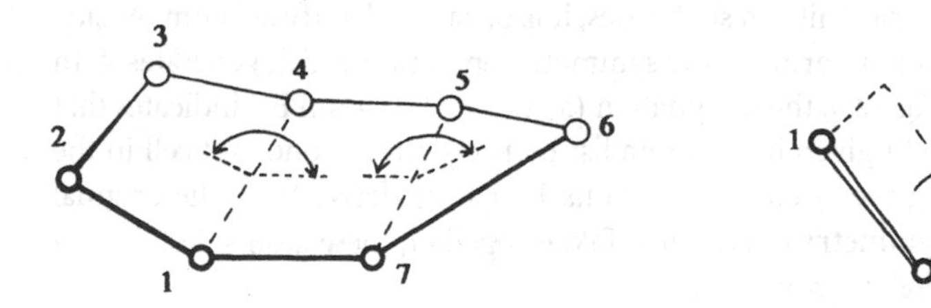

Fig. 12.2. Folding angles in a seven-membered ring.

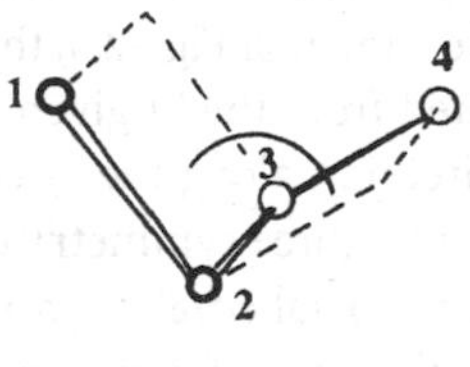

Fig. 12.3. The definition of a torsion angle.

12.3 Structural Geometry and Symmetry

In general, it is not sufficient to discuss only the atoms of the asymmetric unit in a full discussion of the geometry of the structure. In some cases, the application

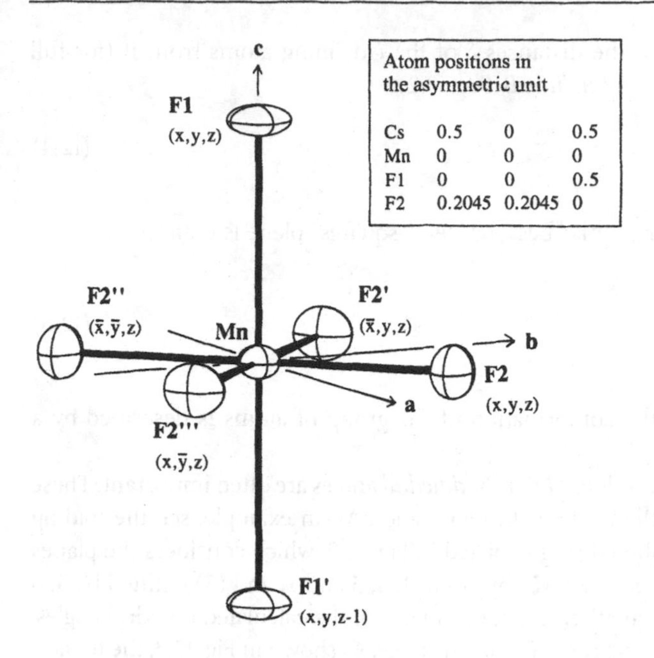

Fig. 12.4. Illustration of symmetry-equivalent positions in an $[MnF_6]^{3-}$ ion in Cs_2MnF_5, space group $P4/mmm$, $a = 6.420$, $c = 4.229$ Å.

of symmetry elements, including lattice translations, may be necessary to construct the complete chemical entity. In most cases, some interesting intermolecular contacts such as hydrogen bonds will only become evident when symmetry related groups are considered. In any case, examining the packing of the crystal requires consideration of all neighboring asymmetric units. In such cases, it is usual to identify an atom outside the basic asymmetric unit in terms of the symmetry operation which generates it. In the example shown in Fig. 12.4, the designation $(x, y, z-1)$ by atom F1′ indicates that it is generated from the F1 given in the atom list by translating it one unit cell in the negative direction along the c-axis. The various F2-atoms derive from the original one (x, y, z) by three symmetry operations. Taken together, these atoms describe an (approximate) octahedral coordination.

Frequently a code for symmetry operations is used which derives from the plotting program ORTEP [58], but is now more generally used in plotting programs. This consists of the number of the symmetry operation (as given in *Int. Tab.* A) followed or preceded by a three digit code for the translation in the ***a***, ***b***, ***c***-directions: 555 implies no translation, 4 implies a translation of -1, 6 a translation of $+1$, 7 of $+2$ etc. The code 75403, for example implies that the atom position is moved by symmetry operation 3, and then translated $+2$ cells in the ***a***-direction and -1 in the ***c***-direction.

In studying coordination geometry, it is very important to know the point symmetry when the central atom occupies a special position (see sect. 6.4). For example, if the Pd atom of a four-fold coordinated Pd atom lies on an inversion center of the space group, it is certain that the complex is planar.

Programs. All of the geometrical calculations mentioned so far are contained in most systems of crystallographic computer programs. A very flexible and independent geometry program which also draws plots is PLATON [70].

12.4 Structural Diagrams

When it comes to drawing illustrations of structures, the problem of exactly what part of the structure is required – which symmetry operations and translations should be included – is once more encountered. Even though the large choice of computer programs available take this into consideration, it is very useful for the beginner to make some plots by hand in order to get a better appreciation of a structure and its symmetry properties. For this purpose, it is best to choose a projection along one of the axes, if possible one perpendicular to the others, e.g. monoclinic ***b***, and to mark the height of each atom in this direction on the plot.

Plotting programs. The various programs may be distinguished by the way in which they select the part of the structure to be drawn and the way in which the atoms and groups of atoms are represented. In the following examples, a few readily

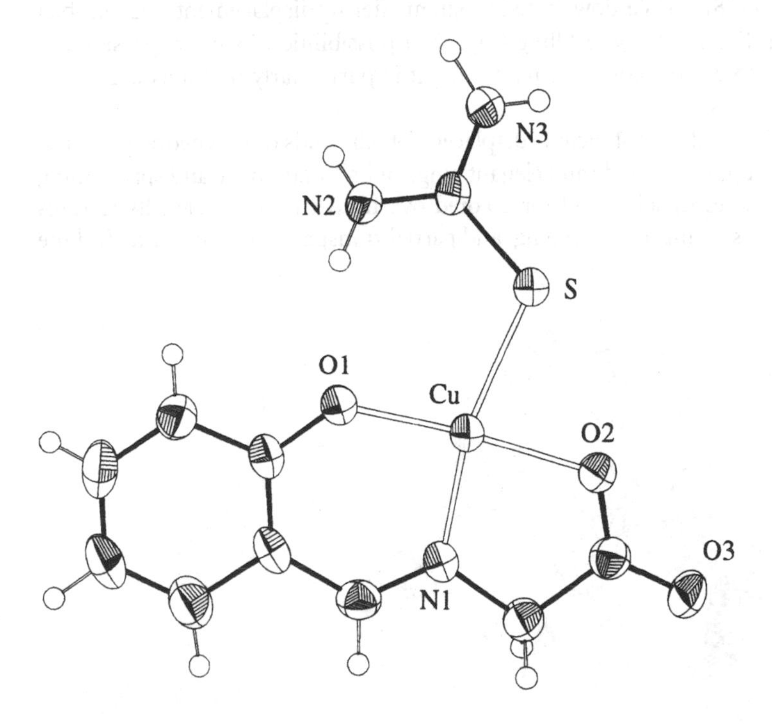

Fig. 12.5. ORTEP-plot [58] of a monomeric unit of the copper complex "CUHABS". Displacement ellipsoids are given at a 50 % probability level, and H-atoms are shown as spheres of arbitrary radius.

available programs are illustrated, and their most characteristic features are noted, for the most part using drawings of the copper complex "CUHABS" (Chapter 15).

In the "classic" *ORTEP*-program [58] (of which interactive versions are available in PLATON [70] and the Bruker-Nonius MaXus System [71]) the atoms can be represented either by their thermal ellipsoids (Fig. 12.5) or as circles, and polyhedra can be shown as "wire-models". Atoms in addition to those in the original asymmetric unit may be generated by the exceptionally elegant "sphere or box" search, and this is particularly useful for structures of non-molecular solids. The instructions that must be provided by the user are rather complex and require some practice.

PLUTO [72] is designed to be a very simple and rapid plotting program. The atoms are represented as shaded spheres – by the use of van der Waals radii, space-filling models may be produced. It is incorporated in the program *PLATON*, mentioned in the last section; it can also draw displacement ellipsoids.

STRUPLO [75] is particularly designed to produce diagrams of inorganic solids represented by coordination polyhedra (Fig. 12.6).

Several programs, including *RASMOL* [84], are available on the Internet. They are particularly useful for displaying and manipulating molecular structures on the screen.

The following programs require licenses or are available commercially:

DIAMOND [78] is a Windows based program offering displacement ellipsoid, ball and stick (e.g. Fig. 12.7), space-filling (Fig. 12.8) possibilities. Its many possibilities allow the best choice of plot to be made, and it is particularly useful as it also has databank functions.

SCHAKAL [73] does not include displacement ellipsoids or polyhedral plots, but provides an exceptionally wide and elegant range of ball-and-spoke and space-filling possibilities. It is especially good for colored overhead transparencies. Its features include variable shading, depth cueing, and partial transparency. In order to find the

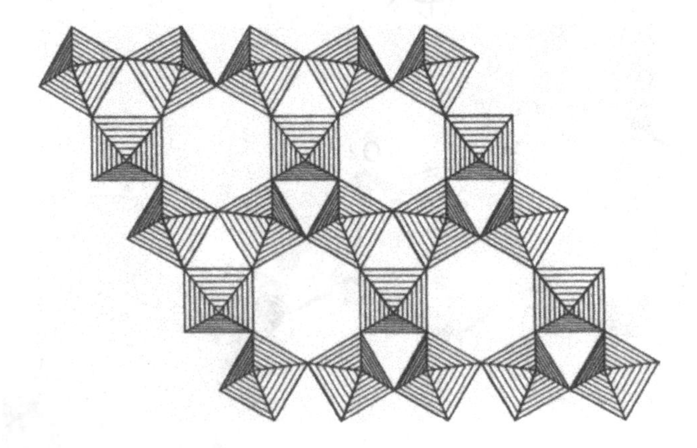

Fig. 12.6. Example of a polyhedral drawing [75] of a layer structure of the Kagomé-type with vertex-sharing octahedra ($Cs_2LiMn_3F_{12}$).

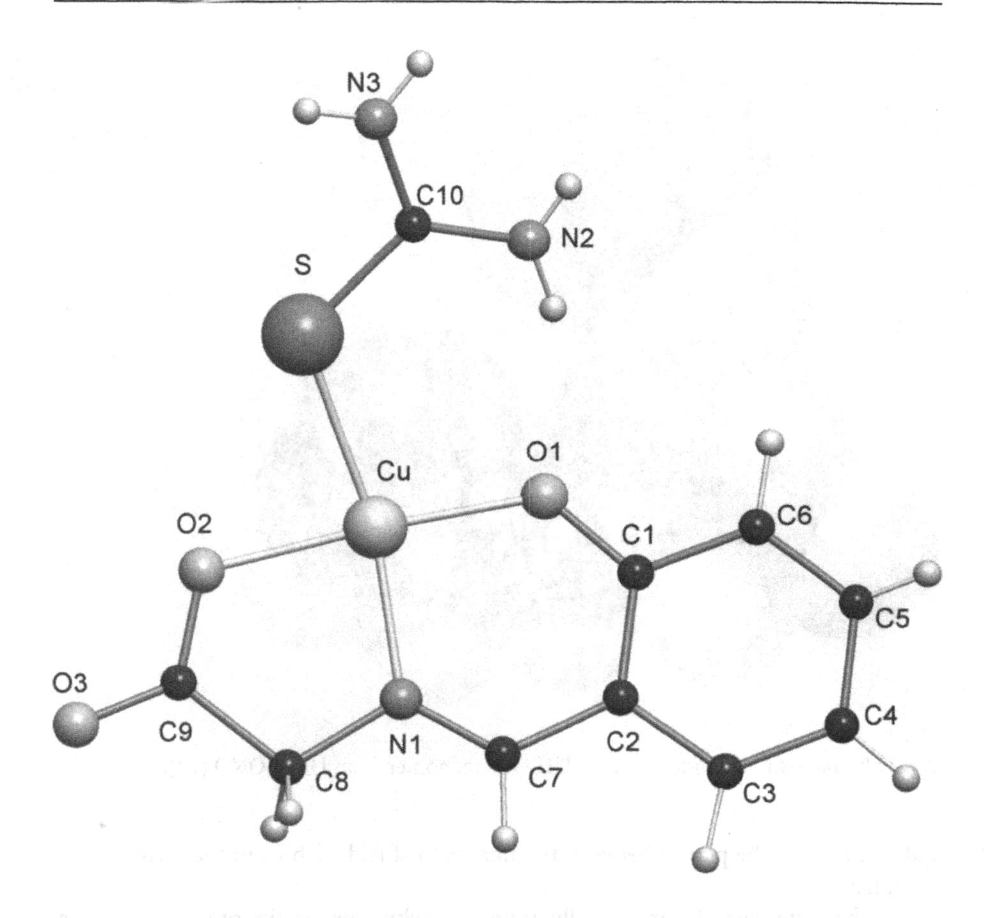

Fig. 12.7. Example of a DIAMOND-plot [78]: ball and stick model of the "CUHABS" monomer.

best view of the structure, a wire model which may be interactively manipulated on the screen is provided. Similar possibilities are also incorporated in other programs, such as *SHELXTL* [74]. See also the list of programs under Molecular Modelling in *ITB* Sect. 3.3.

XPW from *SHELXTL* [74] produces all of types of plots mentioned above, and also gives the possibility of making electron-density plots. It is particularly attractive because of its integration with the *SHELX* system.

ATOMS [80] is another user-friendly program offering a wide range of plots.

Stereo-diagrams. Nearly all of the programs mentioned are capable of producing stereo diagrams. For this purpose, a perspective drawing with a given viewing distance (usually 40–80 cm) must be made. The picture designed for the left eye is then rotated 3° to the right, and that for the right eye 3° to the left. If the two pictures are then placed about 5–6 cm apart, and held at the correct distance from the viewer, the two images can be made to superimpose by relaxing the eyes. If this proves diffi-

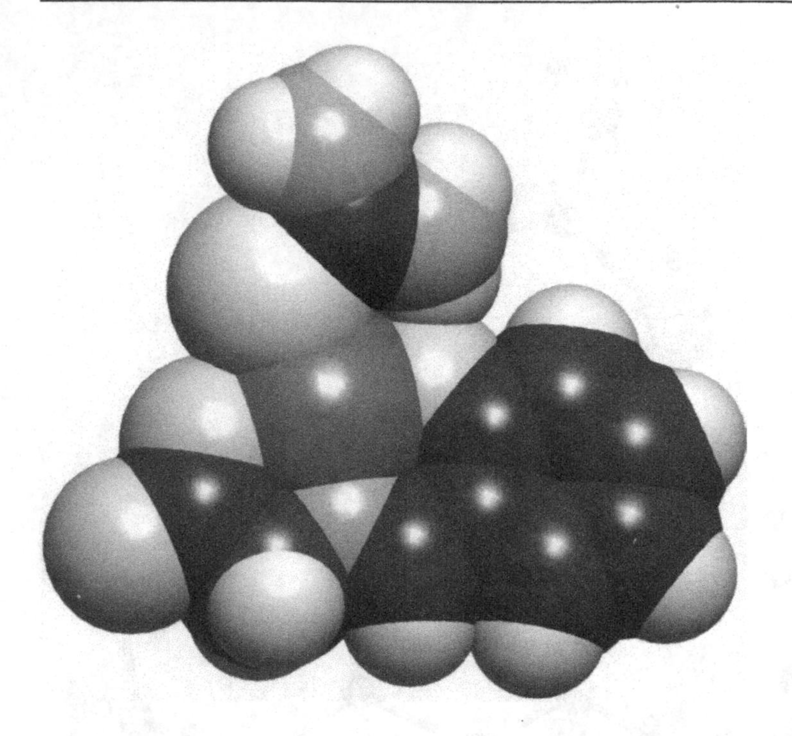

Fig. 12.8. Space-filling model of the "CUHABS" monomer from DIAMOND [78].

cult, a card may be placed between the images to shield each from the "wrong" eye (Fig. 12.9).

Packing diagrams. In order to illustrate the packing properties of the structure, a diagram showing the contents of at least one unit cell is useful (Fig. 12.10).

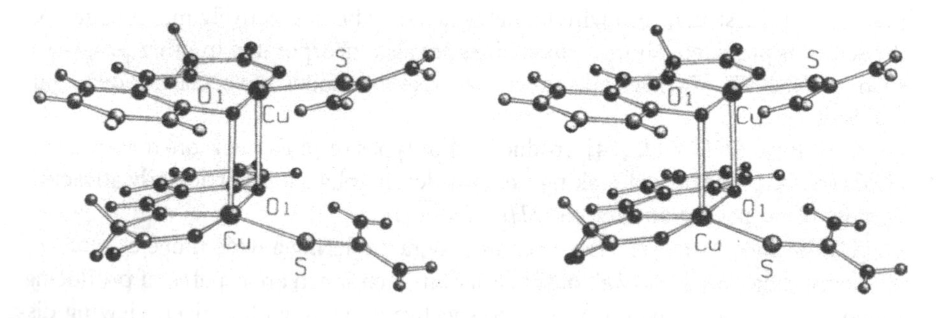

Fig. 12.9. Stereo diagram [73] of the "CUHABS" dimer, generated from the monomer by a 2-fold rotation axis.

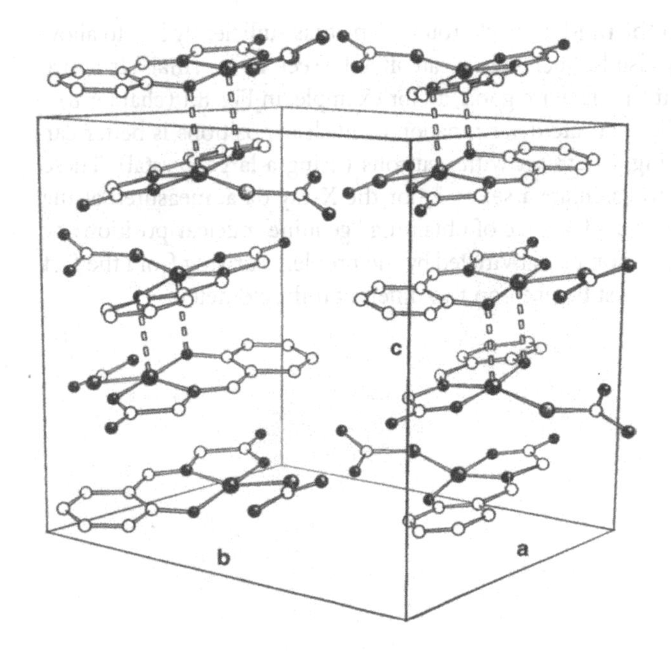

Fig. 12.10. Perspective view of the unit cell of "CUHABS" [73].

12.5 Electron Density

In most cases, the refined positions of the atoms of the structure model are the goal of the structure determination. In some cases however, what is wanted is the direct output of the Fourier synthesis: the actual electron density distribution in the unit cell. With very accurate intensity measurements it is indeed possible to study the fine structure of the electron density: bonding electrons and lone pairs and even the occupation of various orbitals are made "visible" by X-rays.

The X-X Method: Measurements are made at the lowest possible temperature, in order to minimize both atomic vibrations and thermal diffuse scattering, and as accurate a data set as possible is measured to as high a scattering angle as possible. In order to minimise absorption problems, the crystals are often ground to spheres, and all possible interfering effects (see Chapter 11) are avoided or carefully corrected for. Since valence electrons are "smeared" out, they contribute mainly only to reflections with low scattering angle (cf. section 5.1). A refinement based only on *high angle* data will then give the centers of the density of the core-electrons, or essentially the nuclear positions. A structure factor calculation is then carried out using only the spherically symmetric core-electron distribution. When a difference Fourier synthesis is carried out using *all* the diffraction data, the non-spherical part of the electron density, the so-called *deformation density* is emphasized. This is mainly in the vicinity of lone and bonding electron pairs and amounts to a few tenths of an electron per Å^3. The main problem is to ensure that the random background density is sufficiently low to allow these effects to be clearly visible. Such effects can often be seen in "normal" structure determinations already, if the data are good, as, for example, in Fig. 8.1 (chapter 8).

The X-N Method. The accurate determination of nuclear positions is better carried out by first measuring a data set with neutrons (using a large crystal). These positions are then used to calculate a set of F_c for the X-ray data, measured at the same (low) temperature. The advantage of obtaining "genuine" nuclear positions by using neutron diffraction is somewhat vitiated by the problems arising from the fact that two *different* crystals must be used on two different diffractometers.

CHAPTER 13

Crystallographic Databases

The results of crystal structure determinations form the basis of several large databases, which are accessible in most countries of the world through local organisations. Three of these contain most of the data of interest to chemical crystallographers. They may be accessed at national centers, and they are available on CD-ROM.

13.1 The Inorganic Crystal Structure Database (ICSD)

This database contains all structures that have no C—H bonds and are not metals or alloys (i. e. they contain at least one of the elements: H, He, B, C, N, O, F, Ne, Si, P, S, Cl, Ar, Se, Br, Kr, Te, I, Xe, At and Rn. At present, it contains about 70 000 inorganic (i. e. *not* organometallic) structures. It was originally started by G. Bergerhoff of the University of Bonn in 1978, and is currently managed jointly by the NIST (National Institute of Standards and Technology) in the USA, and the FIZ (Fachinformationszentrum Karlsruhe) in Germany. A search with the "FINDIT" software on a Windows platform is started by defining the search criteria, such as chemical elements, a space group, or cell dimensions. The initial output is a list of entries, for each of which the literature reference, the space group, the cell dimensions and the atom parameters may be obtained. Using the menu, further information may be obtained, including bond lengths and angles, and a "visualizer" will give a drawing of the structure, the simulation of its powder diagram or the output of data in cif-format. Figure 13.1 shows the screen input of a search for the elements Na, Mn and F, and a selection of the resulting output.

13.2 The Cambridge Structural Database (CSD)

All crystal structures which *do* contain C—H bonds, that is organic and organometallic structures, (at present more than 290 000, including those in the protein Databank of the Brookhaven National Laboratory) are contained in the CSD, which is maintained by the Cambridge Crystallographic Data Centre (CCDC) in the United Kingdom (www.ccdc.cam.ac.uk). There are mirror sites in several other countries.

Studies are made using the CSD-software CCDC CONQUEST, the key feature of which is the ability to search for structure fragments of given connectivity. A CSD

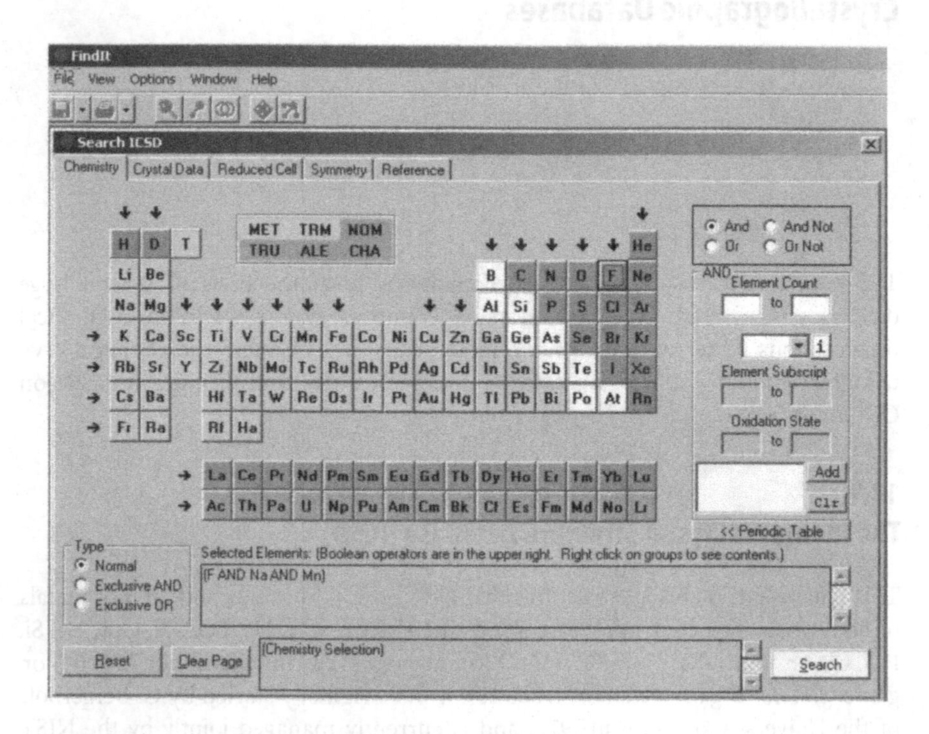

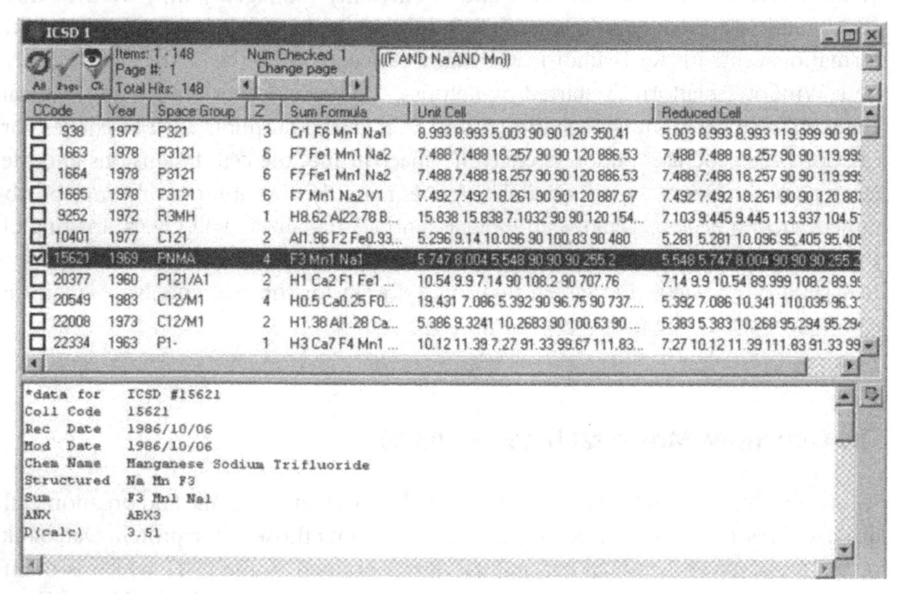

Fig. 13.1. "Findit"-search in the ICSD database for the elements Na, Mn, F, and a selection of the output.

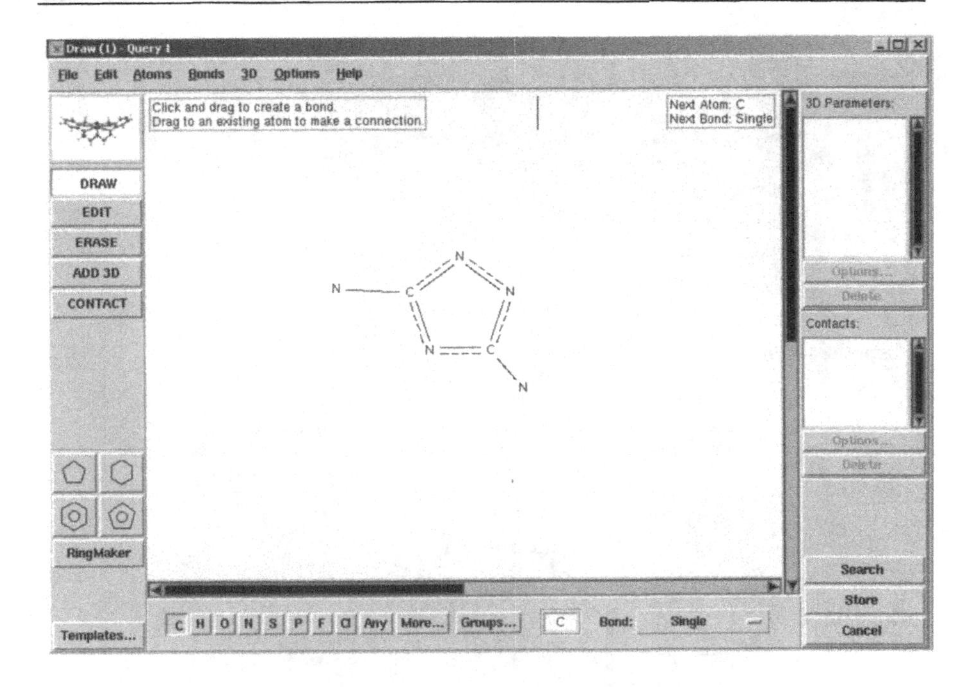

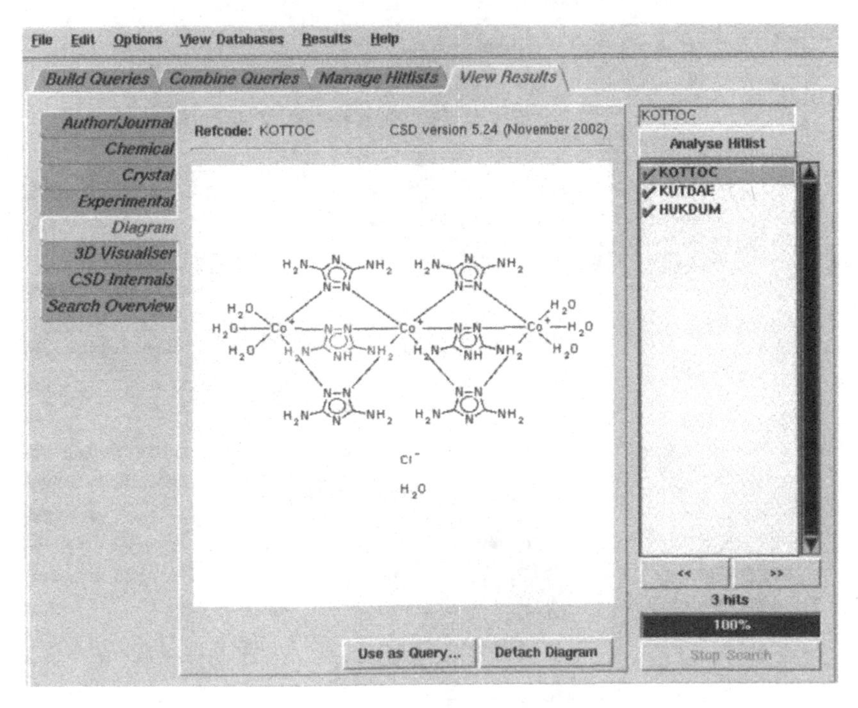

Fig. 13.2. "Conquest"-search in the CSD database of a heterocyclic fragment and selected output for one entry.

Refcode: KOTTOC CSD version 5.24 (November 2002)

Author(s): L.Antolini, A.C.Fabretti, D.Gatteschi, A.Giusti, R.Sessoli

Journal: Inorg.Chem.

Volume: 30

Page: 4858

Year: 1991

Notes:

Deposition Number:

Refcode: KOTTOC CSD version 5.24 (November 2002)

Spacegroup

Name: R-3r **Number:** 148

Cell Parameters

a:	10.333(2)	**b:**	10.333(2)	**c:**	10.333(2)
alpha:	98.33(1)	**beta:**	98.33(1)	**gamma:**	98.33(1)

Volume: 1064.494

Reduced Cell Parameters

a:	10.333	**b:**	10.333	**c:**	10.333
alpha:	98.33	**beta:**	98.33	**gamma:**	98.33

Volume: 1064.494

Other Parameters

Molecular Vol: 1064.494 **Residues:** 3

Z: 1.0 **Z':** 0.17

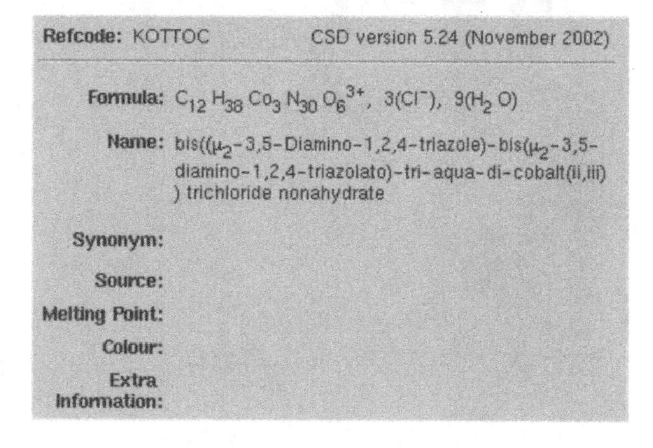

Refcode: KOTTOC CSD version 5.24 (November 2002)

Formula: $C_{12}H_{38}Co_3N_{30}O_6{}^{3+}$, $3(Cl^-)$, $9(H_2O)$

Name: bis((μ_2-3,5-Diamino-1,2,4-triazole)-bis(μ_2-3,5-diamino-1,2,4-triazolato)-tri-aqua-di-cobalt(ii,iii)) trichloride nonahydrate

Synonym:

Source:

Melting Point:

Colour:

Extra Information:

Fig. 13.3. Examples of further output for the selected CSD entry.

search is begun by defining a combination of "queries" relating to the various fields of the database (e.g. chemical composition, elements present, spacegroup, year of publication, authors' names, or the connectivity of the structure, which can be indicated in three dimensions). As the search proceeds, each "hit" produces the literature reference, the main crystallographic data in a variety of formats along with a two-dimensional structure diagram, and usually a "three-dimensional" model, which can be rotated using the mouse. Figure 13.2 & 3 show an example of the input for a search based on a structure-fragment and a selection of the output for one of the "hits".

13.3 The Metals Crystallographic Data File (CRYST-MET)

Metals, alloys, and also semiconductors and borderline cases including metallic phosphides and sulfides (more than 52 000 entries) are stored in the CRYSTMET Database which is maintained at Toth Information Systems in Canada (www.TothCanada.com).

13.4 Other Collections of Crystal Structure Data

As well as the frequent updating of the data bases via the Internet, there are also regular collections of data published in book form. The oldest and best known of these is *Structure Reports* [13], which appeared annually in an organic and an inorganic edition (up to 1990). They give brief accounts of the crystal structures reported during a single year grouped into various classes of compounds. A selection of data from molecular structures in the CSD were published from time to time as *Molecular Structures and Dimensions* [14]. Finally, mention should be made of three older classifications which, because of their systematic coverage are still valuable sources of information [15–17].

13.5 Deposition of Structural Data in Data Bases

When crystal structures are published, the authors are normally requested — or even required — to submit the data, especially material which is not printed, to one or other of the data bases. This means that the full data are available to interested readers of the journal, and the data then become a systematic part of the appropriate data base (ICSD, CSD or CRYST-MET).

CIF-files. The *International Union of Crystallography* (IUCr) has recognized a standard format for the transmission and storage of crystallographic data, called the *crystallographic information file* (CIF). Modern program systems, including [66, 68, 69] provide the possibility for writing these files. It can, usually after a little editing, be sent via the Internet (http://journals.iucr.org/c/services/authorservices.html) or by E-mail directly to the journal or the data base. The journal *Acta Crystallographica*

Section C normally *only* accepts edited CIF-files, and from them prepares the printed article.

13.6 Crystallography on the Internet

A rapidly growing amount of crystallographic information of somewhat variable value is accessible via the Internet. The following are a few of the more interesting addresses:

- `www.iucr.org` (International Union of Crystallography)
- `www.iucr.ac.uk` (Acta Crystallographica)
- `www.unige.ch/crystal/stxnews/stx/welcome.htm` (News group)
- `www.unige.ch/crystal/stxnews/stx/discuss/index.htm` (Discussions)
- `www.ccp14.ac.uk` (Software, etc.)

It is also possible to publish on the Internet. The new section E of Acta Crystallographica (Structure Reports Online) has made it possible for the publication of a "paper" to be achieved in under two weeks from submission.

CHAPTER 14

Outline of a Crystal Structure Determination

The following summary gives an overview of the steps in an X-ray structure determination, together with the chapters and sections in which a more detailed account may be found.

1. Growth of a single crystal (7.1)
2. If required, starting a cooling device (7.1)
3. Choice of a crystal using the polarising microscope (7.1), protecting sensitive compounds with inert oil.
4. Fixing it to the top of a capillary or glass fiber on a goniometer head, and mounting and centering it on the diffactometer. (7.1)

5a. *Measurement using an area-detector system.* (7.3) Recording of a few orientation exposures, peak-search, determination of orientation matrix and unit cell with an indexing program. Choice of final measurement parameters: crystal-to-detector distance, angle width, step width, exposure time. Start of measurement (4–48 hours). Examination of diffraction pattern for superstructure, twinning, satellites, diffuse streaking, determination of reflection profile. Integration of intensities and Lp-correction, accurate determination of cell dimensions.

5b. *Measurement using a four-circle diffractometer.* (7.2.2) Reflection search, possibly using a polaroid photograph, measuring precise goniometer angles for a set of at least 20 reflections. Determination of the orientation matrix and the unit cell with an indexing program, investigation of the reflection profiles. Determination of orientation matrix and unit cell. Possible check of the unit cell with further photographs (7.2.1). Choice of the type and width of the scan, range of scattering angle and of *hkl*-indices, and measuring time (7.2.3). Selection of standard reflections for intensity and orientation control. Automatic data collection (1–14 days) Data reduction and Lp-correction (7.4.1)

6. Possible measurement of the crystal and face-indexing for numerical absorption correction (7.4.3). On a four-circle instrument, measurement of psi-scans.
7. Provisional space group determination (6.6)
8. Structure solution by Patterson (8.2) or direct (8.3) methods.
9. Refinement of the model structure (9.1) and completion of the structure through difference syntheses (8.1). In case of problems, return to 8, 7, 3 or 1 (!)
10. Introduction of anisotropic atomic displacement parameters (5.2) and optimisation of the weighting scheme (9.2)
11. Location or calculation of hydrogen atom positions (9.4.1)
12. Test and possible correction for extinction (10.3) also occasionally for Renninger (10.4) and $\lambda/2$ (10.5) effects

13. Check for alternative space groups, and for non-centric space groups determination of "absolute structure" (10.2)
14. Critical examination of the final model using a checklist such as the following:

   a) Have the "best" cell dimensions been determined? *At the end of the data collection, strong reflections with high scattering angles can be selected to refine the cell parameters. With area-detector systems, very many reflections are used to refine the cell in order to eliminate systematic errors. On four-circle instruments, selected reflections with high 2ϑ-values can be very accurately measured, using the positive and negative settings of the diffractometer. Cell parameter refinement should assume the restrictions of the crystal system – e. g. the 90°-angles of an orthorhombic crystal should not be refined.*

   b) Have the reflection data been correctly used?

      - For F^2 refinement, have all data been used?
      - For F-refinement, has the σ-cutoff been set appropriately?
      - Have reflections been correctly averaged (in particular has the anomalous effect been allowed for in non-centrosymmetric crystals)?
      - Is the number of weak data (e. g. with $F_o < 4\sigma(F_o)$) small enough? If it is high — say > 30 % — is there a good explanation or is an error possible? (see section 11.3–5)
      - Is the distribution of errors uniform over the data set? If there are very many weak data with large errors at high ϑ-values, it may be best to eliminate all data above some ϑ-limit, providing that that does not make the data : parameter ratio too small. (The main concern is not a low R-value but low parameter errors.) Possibly, the weighting scheme should be examined.
      - Has proper correction been made for absorption? Was the correct coefficient μ used? (In the SHELX programs, it is always calculated for the given numbers of atoms on the UNIT instruction)

   c) Is the data : parameter ratio good enough (ideally > 10)

   d) Do the displacement parameters appear reasonable, or are they perhaps masking disorder?

   e) Are the hydrogen positions geometrically sensible and have they been handled in the most appropriate way?

   f) Does the structure display any impossible interatomic contacts?

   g) Is the residual electron density (from a difference Fourier synthesis) reasonably small (largest peaks near heavy atoms)? Is there no evidence of an undetected disordered solvent molecule?

   h) Are the correlations in the refinement small or explainable?

   i) Is the geometry of the structure chemically reasonable?

   j) Are the R-values and the standard deviations of parameters good enough?

k) Has the refinement converged? (i.e. are all the shifts on the last cycle less than 1% of their standard deviations?)

In case of doubt, consideration should be given to the possibility of disorder (10.1), twinning (11.2), wrong unit cell (11.3) or wrong space group (11.4). If so, go back to 5. One of the greatest advantages of area detector systems is that new attempts at cell determination and integration can be made without remeasuring a new crystal, – if the original images have been conserved.

15. Calculation of bond lengths and angles and possibly intermolecular contacts, librational corrections to bond lengths (12.1), selected torsion angles, "best" planes (12.2) Preparation of tables for publication and deposition in databanks.
16. Examination of crystal packing (12.3) and design of structure diagrams (12.4)
17. Explanation of the structure, discussion of its significance and comparison with others.

CHAPTER 15

Worked Example of a Structure Determination

In this chapter, the most important stages in a structure determination will be given using as an example the thiourea adduct of N-salicylideneglycinatocopper(II) $Cu(sg)SC(NH_2)_2$, ("CUHABS," molecular formula $C_{10}H_{11}N_3O_3SCu$, C. Friebel, Marburg: cf. Fig. 12.5 and 12.7–12.9). Computer output is given in an abbreviated form and commentated in italic.

1. *Selection of a crystal (ca. 0.3 × 0.2 × 0.1 mm) using a polarising microscope and fixing it on the top of an open quartz capillary on a goniometer head using a perfluorinated grease. Mounting it on an IPDS area detector system, centering and height adjustment, starting the cooling device, cooling down to 193 K. Recording of three orienting images with graphite monochromatized Mo radiation at φ intervals of 0–1.2° (Fig. 15.1), 1.2–2.4°, 2.4–3.6°.*
2. *Check of reflection profiles and intensities: preliminary indexing gives a monoclinic I-centered cell with $a = 13.64(2)$, $b = 12.37(1)$, $c = 14.21(1)$ Å, $\beta = 91.3(1)°$. (The "standard" C-setting in the monoclinic system would give β a minimum value of 131.7°.) Based on this, the final measuring conditions were fixed: Distance crystal to image plate: 60 mm (for a plate diameter of 180 mm, a ϑ range to 28.2° is then accessible), measuring time 5 min per record, 167 records from $\varphi = 0°$ to 200° in steps of 1.2°.*
3. *After a measuring time of 1 day, the orientation matrix and thus the cell dimensions were determined from the peaks of 40 records:*

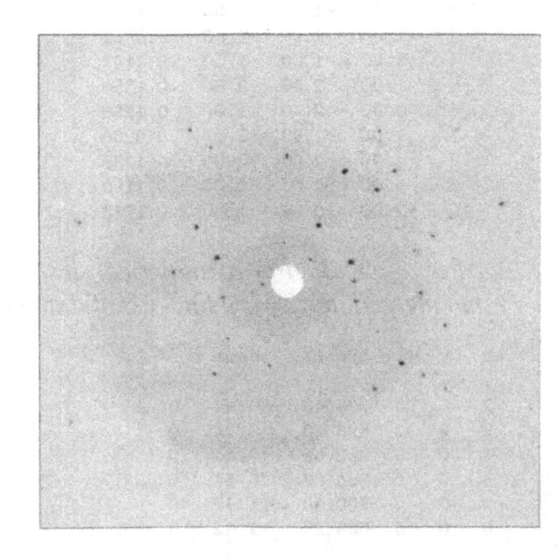

Fig. 15.1. First area-detector exposure (section $f = 0$–1.2°).

```
Reciprocal axis matrix
                          0.017251 -0.030457   0.062903
                          0.047297  0.060659   0.009800
                         -0.053275  0.043998   0.030032
```

4. *Determination of the 3D reflection profile (with the help of a program) and integration of all reflections, i. e. evaluation of gross intensities, subtraction of background, and application of Lorentz- and polarisation corrections. The resulting data file CUHABS.RAW (sample below) contains a total of 11554 reflections. For each reflection are given the hkl indices, the net intensities and their estimated errors, and the six direction cosines which give the orientation of the incident and reflected beams for that reflection with respect to the reciprocal axes.*

```
-14   2   2  645.12 11.60 2 -0.55864 -0.24445 -0.39613  0.49938 -0.73713 0.82406
-13   4   1  727.14 11.45 2 -0.55864 -0.18825 -0.39613  0.59745 -0.73713 0.77306
-13   8  -3   86.41  8.48 2 -0.55864 -0.19349 -0.39613  0.79796 -0.73713 0.56373
-12  10  -6    8.80  9.22 2 -0.55864 -0.14021 -0.39613  0.89949 -0.73713 0.40748
-10 -11   5   60.98  8.10 2 -0.55864 -0.01089 -0.39613 -0.14807 -0.73713 0.98918
 -9 -15   4   63.37  9.00 2 -0.55864  0.04643 -0.39613 -0.34798 -0.73713 0.93881
 -7   9  -4  247.26  8.41 2 -0.55864  0.14989 -0.39613  0.84304 -0.73713 0.51472
 -7  11  -8   26.56  8.08 2 -0.55864  0.14516 -0.39613  0.94030 -0.73713 0.30488
 -6  11  -9  129.93  8.40 2 -0.55864  0.20304 -0.39613  0.94486 -0.73713 0.25426
 -5  10  -7 1319.86 12.37 2 -0.55864  0.26383 -0.39613  0.89455 -0.73713 0.35985
......
```

5. *From the entire data set, 5 000 stronger reflections were selected by a program to refine the final accurate cell parameters:*

```
Cell Parameters
    12.3543(12)        14.3081(14)       13.5164(12)
    90.0               91.352(12)        90.0
```
a, b, c [Å]
α, β, γ [°]

6. *All visible faces of the crystal were in turn aligned parallel to the view direction (with the aid of a CCD camera) and their hkl indices and distances to the center of the crystal are determined for a numerical absorption correction.*

```
Crystal Faces

     N      H       K       L      D [mm]
     1    1.00   -1.00    1.00    0.1110
     2   -1.00   -1.00    2.00    0.1419
     3    1.00    1.00   -2.00    0.0952
     4   -1.00    1.00   -1.00    0.1148
     5    1.00    2.00    0.00    0.1350
     6    0.00    2.00    1.00    0.1256
     7    1.00   -1.00   -2.00    0.0960
     8   -1.00   -2.00    0.00    0.1299
     9    0.00   -2.00   -1.00    0.1160
    10   -2.00    0.00   -1.00    0.1613
```

7. *After application of a numerical absorption correction, the data were sorted but not merged, to give the final list of data for solution and refinement of the structure.*

```
   h   k   l   Fo**2    sigma
.........
  -2   0   0 1846.40    30.47   0
   2   0   0 1826.90    30.16   0
  -4   0   0  262.59     9.54   0
   4   0   0  264.68    10.67   0
  -6   0   0  700.48    11.47   0
   6   0   0  685.38    12.42   0
```

```
   8   0   0 1787.91    17.01   0
  -8   0   0 1838.32    16.57   0
.........
  -3   0  -1     2.00    6.33   0
   3   0   1     2.88    5.39   0
  -5   0  -1     0.45    4.96   0
   5   0   1     1.82    9.27   0
  -7   0  -1    -2.42    4.43   0
   7   0   1    -2.11    6.04   0
.........
  -4   0  -2 6527.41    31.42   0
   4   0   2 6679.64    32.72   0
  -6   0  -2 6602.95    28.64   0
   6   0   2 6682.05    29.15   0
  -8   0  -2  210.17     7.23   0
   8   0   2  211.58     8.77   0
```

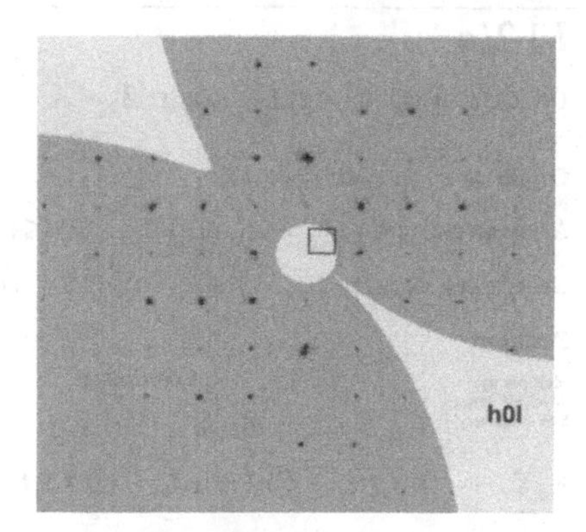

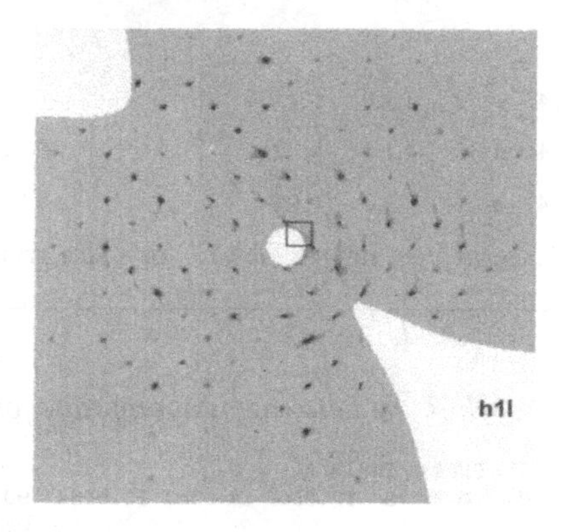

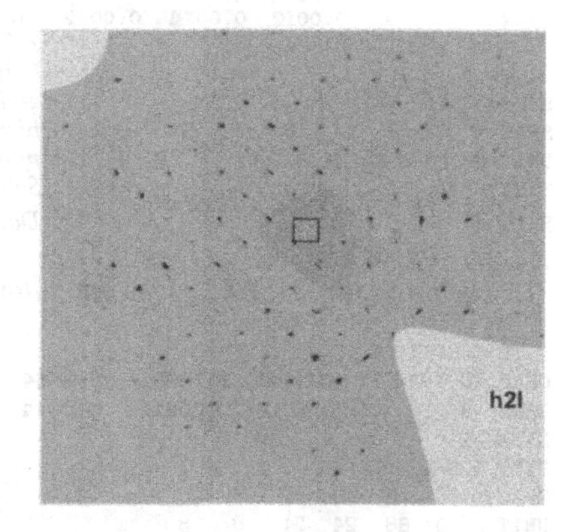

Fig. 15.2. Layers $h0l$, $h1l$ and $h2l$ of the reciprocal lattice.

8. *Search for systematic absences in the data set. The data measured include none violating the general condition for an I-lattice (hkl: $h + k + l = 2n$). In addition, data for the zone h0l are very weak when $h \neq 2n$ (see sample above). This implies an a-glide plane normal to the b-axis. These conditions indicate space group I2/a (15) or Ia (9). This is most convincingly seen when the measured data are displayed as layers of the reciprocal lattice. (Fig. 15.2)*
9. *Estimation of Z, the number of molecules per unit cell. The molecular formula for the complex:* $C_{10}H_{11}N_3O_3SCu$ *contains 18 non-hydrogen atoms, so the molecular volume should be about* $18 \times 17 = 306$ Å^3, *and the unit cell volume of: 2389* Å^3 *is thus compatible with* $Z = 8$. *The general position in space group I2/a is 8-fold (in Ia only 4-fold), i. e. in I2/a the complex molecule need have no crystallographic symmetry imposed on it. The first attempt at a structure solution will thus assume the centrosymmetric space group I2/a. Note that the* $E^2 - 1$ *test (see below) also suggests a centrosymmetric space group.*
10. *Selection of a method of solution. The heaviest atom, Cu, has 18 % of the electrons in the formula, making this structure suitable for either Patterson or direct methods. Both possibilities will be illustrated here using the program SHELXS-97 [64]. Both methods read the reflection data file CUHABS.HKL and a file CUHABS. INS, of*

$I\,1\,2/a\,1$

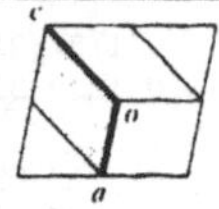

UNIQUE AXIS b, CELL CHOICE 3

Origin at $\bar{1}$ on glide plane a

Asymmetric unit $0\le x\le 1;\ 0\le y\le \frac{1}{4};\ 0\le z\le \frac{1}{2}$

Generators selected (1); $t(1,0,0)$; $t(0,1,0)$; $t(0,0,1)$; $t(\frac{1}{2},\frac{1}{2},\frac{1}{2})$; (2); (3)

Positions

Multiplicity, Wyckoff letter, Site symmetry	Coordinates	Reflection conditions
	$(0,0,0)+\quad(\frac{1}{2},\frac{1}{2},\frac{1}{2})+$	General:
8 f 1	(1) x,y,z (2) $\bar{x}+\frac{1}{2},y,\bar{z}$ (3) $\bar{x},\bar{y},\bar{z}$ (4) $x+\frac{1}{2},\bar{y},z$	$hkl: h+k+l=2n$; $h0l: h,l=2n$; $0kl: k+l=2n$; $hk0: h+k=2n$; $0k0: k=2n$; $h00: h=2n$; $00l: l=2n$
		Special: as above, plus
4 e 2	$\frac{1}{4},y,0\quad \frac{3}{4},\bar{y},0$	no extra conditions
4 d $\bar{1}$	$\frac{1}{4},\frac{1}{4},\frac{3}{4}\quad \frac{3}{4},\frac{1}{4},\frac{3}{4}$	
4 c $\bar{1}$	$\frac{1}{4},\frac{1}{4},\frac{1}{4}\quad \frac{3}{4},\frac{1}{4},\frac{1}{4}$	$hkl: l=2n$
4 b $\bar{1}$	$0,\frac{1}{2},0\quad \frac{1}{2},\frac{1}{2},0$	
4 a $\bar{1}$	$0,0,0\quad \frac{1}{2},0,0$	$hkl: h=2n$

Excerpt from Intern. Tables A for space group $I2/a$.

which the following is appropriate for a direct methods solution

```
TITL CUHABS IN I2/A
CELL  0.71073  12.3543 14.3081 13.5164  90.00  91.352  90.00
                                        Wavelength of X-rays and cell dimensions
ZERR  8            0.0012  0.0014  0.0012   0.0    0.012    0.0
                                        Z (formulae per unit cell) and errors on cell dimensions
LATT  2                                 Lattice I - centrosymmetric space group
SYMM 0.5+X,-Y,Z                         a-glide plane normal to y — other symmetry implied
SFAC    C   H  CU   O   N   S           Element types present in unit cell
UNIT 80 88 8 24 24 8                    number of atoms of each type per unit cell
TREF                                    Solution by direct methods — default options to be used
HKLF 4                                  Data as h, k, l, F^2 to be read
```

11. *The program SHELXS gives the following (abridged) on the file CUHABS.LST.*

```
TITL CUHABS IN I2/A
CELL  0.71073   12.3543  14.3081  13.5164  90  91.352  90
ZERR  8          0.0012   0.0014   0.0014   0   0.012   0
LATT  2
SYMM 0.5+X,-Y,Z
SFAC    C   H   O   N   S  CU
UNIT   80  88  24  24   8   8
                                            Check data calculated each run by program
V =      2388.58   At vol =     16.6      F(000) =  1288.0         mu = 2.01 mm-1
Max single Patterson vector =  58.1      cell wt =    2534.54    rho =  1.742
TREF
HKLF 4
    11554 Reflections read, of which 179 rejected
                                        11375 data are available omitting systematic absences
    Maximum h, k, l and 2-Theta =    16.   18.   17.   55.99
INCONSISTENT EQUIVALENTS             Data merging — equivalents with large discrepancies flagged
  h   k   l        F*F     Sigma(F*F)  Esd of mean(F*F)
  1   1   0     244.44        4.42     172.11
```

```
  1   7   0      6708.11       12.09       83.27
  7   7   0      3528.96        8.82       67.27
  1  13   0      3743.91        7.62       79.39
** Etc. **
 2755  Unique reflections, of which   2381  observed
```

Most of the data have equivalents — 2381 unique data have $F^2 > 4\sigma$

```
 R(int) = 0.0345      R(sigma) = 0.0243        Friedel opposites merged
```

R(int) based on consistency of merged data, R(sigma) based on esd's from diffractometer

```
 Observed E .GT. 1.200 1.300 1.400 1.500 1.600 1.700 1.800 1.900 2.000 2.100
 Number             758   673   589   513   437   369   314   252   200   163
                    Centric Acentric     0kl        h0l        hk0       Rest
 Mean Abs(E*E-1)     0.968     0.736     0.922      0.896      0.996      0.965
```

These tests are consistent with the choice of the centrosymmetric space group I2/a rather than the acentric Ia

```
SUMMARY OF PARAMETERS FOR  CUHABS IN I2/A
```

Default values for parameters generated by the instruction "TREF". Note particularly np = number of solutions attempted; nE = number of strong E-values to be used for direct methods.

```
SUMMARY OF PARAMETERS FOR CUHABS IN I2/A
 ESEL  Emin   1.200     Emax  5.000     DelU 0.005     renorm 0.700     axis 0
 OMIT  s  4.00     2theta(lim)  180.0
 INIT  nn    11     nf    16     s+  0.800     s-  0.200     wr   0.200
 PHAN  steps  10  cool 0.900   Boltz 0.400  ns 146  mtpr   40     mnqr   10
 TREF  np       256.     nE    232     kapscal  0.900      ntan    2     wn -0.950
 FMAP  code  8
 PLAN  npeaks    -25      del1 0.500      del2 1.500
 MORE  verbosity  1
 TIME  t       9999999.
     146 Reflections and    1641. unique TPR for phase annealing
     232 Phases refined using      5524. unique TPR
     320 Reflections and     8741. unique TPR for R(alpha)
```

A subset of the full list of Triple Product Relationships (TPR) are used for phase annealing. Eventually, 232 phases will be refined and 320 used for a consistency test

```
  892 Unique negative quartets found, 892 used for phase refinement
ONE-PHASE SEMINVARIANTS
```

These are the data with potential Σ_1*-relationships (usually data with h, k and l all even.) There are 65 in the list of 234 data to be phased.*

```
    h    k    l       E        P+      Phi
    0    6    4    2.716
    2    6    0    2.446     0.39
   -8    6    4    2.510
   -2    6    4    2.375     0.39
    6    6    2    2.417
......
Expected value of Sigma-1 = 0.888
Following phases held constant with unit weights for the initial 4 weighted
tangent cycles (before phase annealing):
```

Phases chosen for starting set for phase expansion

```
    h    k    l       E        Phase/Comment

    0    5    1    2.222     random phase
    4    0    2    1.728     180      sigma-1 = 0.118
    4    0    6    1.801       0      sigma-1 = 0.871
    5    2    1    2.094     random phase
    0    5    5    2.108     random phase
    1    5    0    2.059     random phase
.......
All other phases random with initial weights of 0.200 replaced by 0.2*alpha
(or 1 if less) during first 4 cycles - unit weights for all phases thereafter
      124 Unique NQR employed in phase annealing
      128 Parallel refinements

STRUCTURE SOLUTION for  CUHABS IN I2/A
Phase annealing cycle:      1    Beta =   0.04571
```

Initial treatment of a few phases to help avoid false minima

```
Ralpha 1.438 0.623 0.327 1.360 0.605 0.032 3.064 0.051 0.392 0.334 1.077 0.027 .....
Nqual  0.089-0.311-0.504-0.085-0.595-0.781-0.019-0.946-0.784-0.551 0.091-0.951 .....
Mabs   0.443 0.585 0.701 0.452 0.591 1.181 0.338 1.047 0.673 0.691 0.488 1.130 .....
......
```

The following (abbreviated!) list gives figures of merit and signs for the 65 seminvariants for each of the 256 solutions. If these differ between two sets, they are probably different solutions. The correct solution normally has the minimum value of CFOM.

```
Try    Ralpha  Nqual  Sigma-1  M(abs) CFOM Seminvariants

  733881. 0.061 -0.169 0.707 1.317 0.671  +-+-- ++++- +-+++ +---+ +++++ --+++ --++- .....
  384789. 0.061 -0.169 0.707 1.317 0.671  +-+-- ++++- +-+++ +---+ +++++ --+++ --++- .....
 1109605. 0.060 -0.975 0.912 1.308 0.060* --+++ -+++- -+--- +-+++ +---- --+-+ +--++ .....
 2055245. 0.060 -0.975 0.912 1.308 0.060  --+++ -+++- -+--- +-+++ +---- --+-+ +--++ .....
 1887617. 0.060 -0.975 0.912 1.308 0.060  --+++ -+++- -+--- +-+++ +---- --+-+ +--++ .....
 1161201. 0.087 -0.990 0.649 1.053 0.087  ---+- ++--+ +-+-- +++-- ++++- +-++- -++-- .....
.....
CFOM Range   Frequency
 0.000 -0.020     0
 0.020 -0.040     0
 0.040 -0.060    90
 0.060 -0.080     0
 0.080 -0.100    43
 0.100 -0.120     0
 0.120 -0.140     0
 0.140 -0.160     0
.....
```

Indicating that the structure is (probably) solved!

```
256. Phase sets refined - best is code  1109605.  with CFOM =  0.0599
 Tangent expanded to  758  out of  758  E greater than  1.200
 FMAP and GRID set by program
 FMAP    8   3  17
 GRID    -1.786  -2  -1     1.786   2   1
 E-Fourier for CUHABS IN I2/A
 Maximum = 620.68, minimum = -132.67 highest memory used = 8780 / 13196
 Heavy-atom assignments:
          x        y        z     s.o.f.    Height
 CU1    0.3055  0.3267  0.1193  1.0000    620.7
 S2     0.3381  0.4802  0.1480  1.0000    315.4
```

Interpretation of strongest peaks in E-map — they look good — there are as many Cu and S atoms as expected

```
 Peak list optimization
        RE = 0.162 for  16 surviving atoms and  758 E-values
 E-Fourier for  CUHABS IN I2/A
        Maximum =  612.55,  minimum = -142.04
 Peak list optimization
        RE = 0.135 for  18 surviving atoms and  758 E-values
 E-Fourier for  CUHABS IN I2/A
        Maximum =  616.95,  minimum = -106.18 0.4 seconds elapsed time
```

After two cycles of peak optimisation, the R-value is satisfactory

```
Molecule  1     scale 0.786 inches = 1.997 cm per Angstrom
```

"Lineprinter" plot of results (scale refers to plot using 12 point Courier font). The plot contains some atoms symmetry related to the basic set found. In this example, they are not needed for the interpretation. The program has "found" the Cu and S atoms — the rest of the interpretation is by the crystallographer! Note that the solution is not perfect — peaks 8, 9, 10, 12 and 18 all represent noise, and must be discarded on chemical grounds. This hand-interpretation of the peaks and the build-up of a model is in practice now most often done of the computer screen using a graphics program such as SHELXTL [74] or WINGX [81].

The assignment of chemically sensible atom-types is assisted by the following list of distances and angles. The list has been shortened by the removal of some peaks which are noise - peak 18, and all beyond 21.

```
Atom Peak     x        y        z      SOF  Height  Distances and Angles
CU1  0.   0.3055  0.3267  0.1193  1.000  3.08    0 S2    2.264
```

Distance Cu1–S2

```
                                                 0 1     1.967  80.9
```

Distance Cu1–1 and angle S2–Cu1–1

```
                                                 0 2     1.911 103.0 174.1
```

Distance Cu1–2 and angles S2–Cu1–2, 1–Cu1–2

```
                                                 0 10    2.251  29.9 104.6  77.8      etc
                                                 0 11    2.017 159.8  84.2  93.0 170.1
 S2   0.   0.3381  0.4802  0.1480  1.000  3.03    0 CU1   2.264
                                                 0 4     1.782 116.0
                                                 0 10    1.164  74.4  61.6
 1  192.   0.4622  0.3202  0.1483  1.000  2.95    0 CU1   1.967                  O2
                                                 0 7     1.308 115.4
```

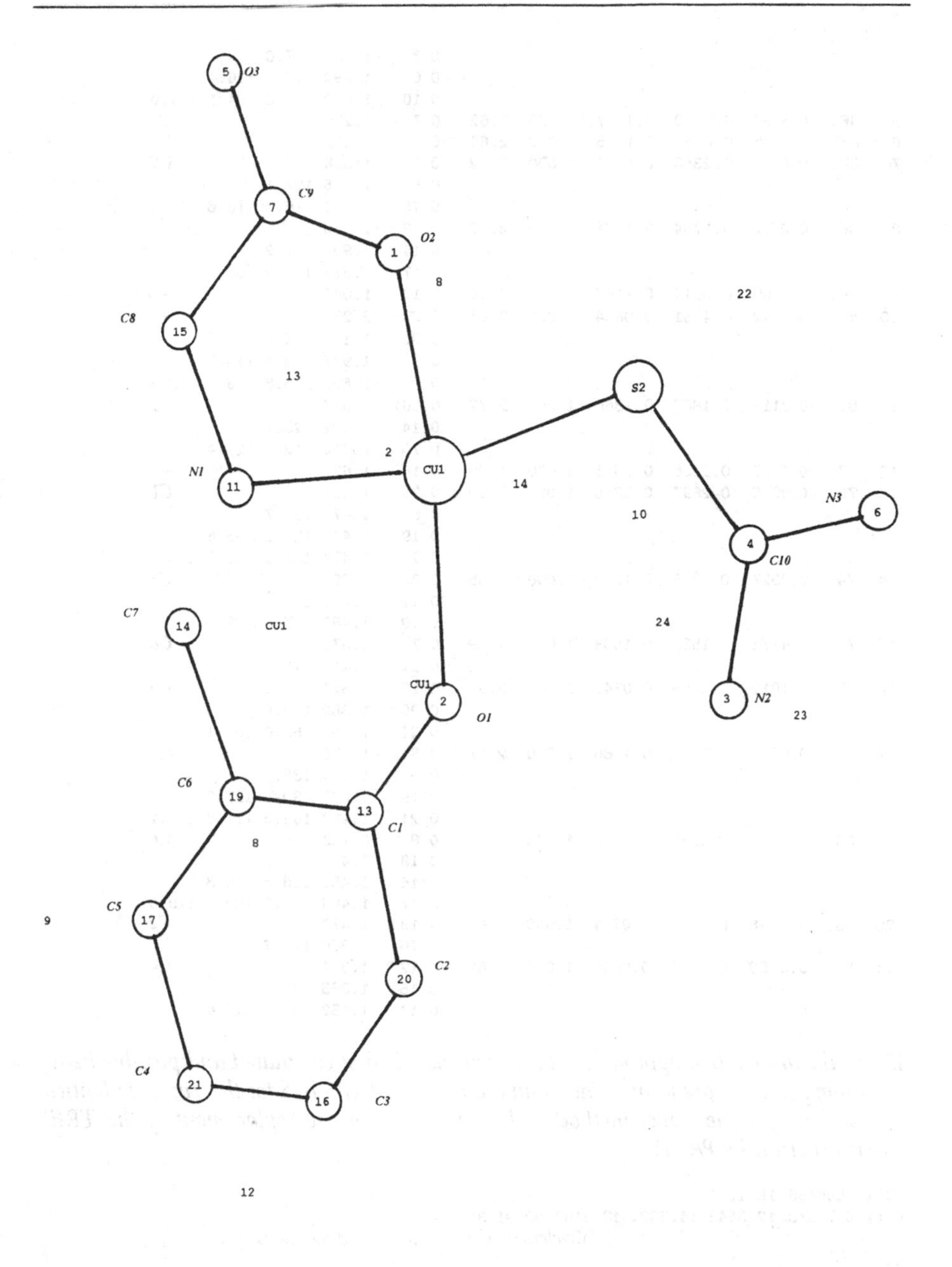

2	189.	0.1562	0.3254	0.0794	1.000	3.33	0	CU1	1.911		O1
							0	13	1.221	126.7	
3	105.	0.1278	0.5205	0.1189	1.000	3.14	0	4	1.337		N2
							0	10	1.977	53.6	
4	100.	0.2219	0.5533	0.1557	1.000	2.91	0	S2	1.782		C10

```
                                                               0 3      1.337 117.8
                                                               0 6      1.292 115.7 126.4
                                                               0 10     1.600  39.8  84.1 142.0
5   96.  0.5993  0.2140  0.1817  1.000  2.62                   0 7      1.275                 O3
6   87.  0.2385  0.6351  0.1935  1.000  2.64                   0 4      1.292                 N3
7   87.  0.5008  0.2363  0.1640  1.000  2.72                   0 1      1.308                 C9
                                                               0 5      1.275 127.1
                                                               0 15     1.571 116.2 116.6
8   86.  0.0792  0.1754  0.1471  1.000  2.32                   0 13     1.478                 -
                                                               0 14     1.995  91.2
                                                               0 17     1.376 122.2  93.3
9   86.  0.0509  0.0214  0.1167  1.000  2.40                   0 17     1.044                 -
10  84.  0.2742  0.4781  0.0834  1.000  3.63                   0 CU1    2.251                 -
                                                               0 S2     1.164  75.7
                                                               0 3      1.977 113.4 114.8
                                                               0 4      1.600 125.8  78.5  42.3
11  80.  0.3113  0.1867  0.1360  1.000  2.72                   0 CU1    2.017                 N1
                                                               0 14     1.202 123.1
                                                               0 15     1.410 112.5 124.4
12  77. -0.1597  0.1356  0.1511  1.000  1.95                   0 16     1.627                 -
13  75.  0.0936  0.2597  0.0856  1.000  3.10                   0 2      1.221                 C1
                                                               0 8      1.478 138.7
                                                               0 19     1.442 128.2  28.6
                                                               0 20     1.477 121.3  91.3 110.3
14  74.  0.2327  0.1375  0.1280  1.000  2.65                   0 8      1.995                 C7
                                                               0 11     1.202 126.7
                                                               0 19     1.480  17.0 129.3
15  72.  0.4171  0.1539  0.1563  1.000  2.59                   0 7      1.571                 C8
                                                               0 11     1.410 111.6
16  70. -0.1012  0.2058  0.0741  1.000  2.92                   0 12     1.627                 C3
                                                               0 20     1.350 138.4
                                                               0 21     1.355  53.0 121.1
17  64.  0.0386  0.0932  0.1086  1.000  2.57                   0 8      1.376                 C5
                                                               0 9      1.044 138.7
                                                               0 19     1.408  30.1 127.2
                                                               0 21     1.452 102.3 111.6 121.2
19  59.  0.1180  0.1638  0.1099  1.000  2.74                   0 8      0.723                 C6
                                                               0 13     1.442  78.5
                                                               0 14     1.480 126.3 118.3
                                                               0 17     1.408  72.5 122.5 119.0
20  55. -0.0248  0.2733  0.0764  1.000  3.08                   0 13     1.477                 C2
                                                               0 16     1.350 126.7
21  54. -0.0752  0.1154  0.0928  1.000  2.63                   0 12     1.354                 C4
                                                               0 16     1.355  73.8
                                                               0 17     1.452 135.9 117.4
```

12. *In the alternative approach, a Patterson map is first calculated and possible heavy atom positions are found. The input data on CUHABS.INS for this run are identical to those for the direct methods solution except for the replacement of the TREF instruction by PATT:*

```
TITL  CUHABS IN I2/A
CELL 0.71073 12.3543 14.3081 13.5164 90 91.352 90
                                      Wavelength of X-rays and cell dimensions
.....
UNIT 80 88 8 24 24 8                  number of atoms of each type per unit cell
PATT                                  Solution by Patterson methods — default options to be used
HKLF 4                                Data as h, k, l, F^2
```

13. *The output produced on CUHABS.LST is now:*

```
TITL  CUHABS IN I2/A
CELL  0.71073   12.3543  14.3081  13.5164  90  91.352  90
ZERR  8          0.0012   0.0014   0.0014   0   0.012   0
```

```
LATT  2
SYMM  0.5+X, -Y, Z
SFAC    C   H   N   O   S  CU
UNIT   80  88  24  24   8   8
......
    SUMMARY OF PARAMETERS FOR CUHABS IN I2/A
```

Default values for parameters for Patterson calculation and search. Note particularly nv = 1 — only best solution to be attempted. This can be increased if first solution unsuitable.

```
ESEL  Emin  1.200    Emax  5.000    DelU 0.005     renorm 0.700     axis 0
OMIT  s  4.00    2theta(lim)  180.0
PATT  nv   1   dmin    1.80   resl  0.76   Nsup  206    Zmin  5.80    maxat  8
FMAP  code  6
PLAN  npeaks    80    del1 0.500     del2 1.500
MORE  verbosity  1
TIME  t     9999999.
FMAP and GRID set by program
```

Map chosen to cover one asymmetric unit (in I2/m for vector map) with a grid of about 0.25 Å.

```
FMAP   6   3  17
GRID    -1.786 -2  -1      1.786   2   1
Super-sharp Patterson for CUHABS IN I2/A
```

Vector map — some of the peaks have been interpreted in terms of eventual solution. Relative peak heights and distance from origin given.

```
Maximum = 999.10, minimum = -140.65 highest memory used = 9228 / 16556
```

	X	Y	Z	Weight	Peak	Sigma	Length	
1	0.0000	0.0000	0.0000	4.	999.	77.53	0.00	*The origin peak — given as 999 to scale others*
2	0.0000	0.8468	0.5000	2.	323.	25.04	7.10	*The "Harker line" peak for copper* $(0, 1\frac{1}{2} - 2y, \frac{1}{2})$
3	0.1142	0.0000	0.2349	2.	305.	23.65	3.44	*The "Harker plane" peak for copper* $(-\frac{1}{2} + 2x, 0, 2z)$
4	0.1521	0.0000	0.0331	2.	176.	13.64	1.92	*Cu—N and Cu—O bond vectors superimposed*
5	0.1139	0.8472	0.7349	1.	145.	11.27	4.45	*Cu—Cu "general vector"* $(-\frac{1}{2} + 2x, 1\frac{1}{2} - 2y, \frac{1}{2} + 2z)$
6	0.8590	0.8472	0.7341	1.	139.	10.80	4.52	
7	0.5318	0.1926	0.0302	1.	137.	10.65	6.43	
8	0.8556	0.3075	0.2322	1.	135.	10.46	5.71	
9	0.0318	0.1558	0.0301	1.	112.	8.67	2.30	*Cu—S bond vector*

```
......
Patterson vector superposition minimum function for CUHABS IN I2/A
Patt. sup. on vector 1 0.0000 0.8468 0.5000 Height 323. Length 7.10
Maximum = 287.22, minimum = -127.35 highest memory used = 13037 / 32121
68 Superposition peaks employed, maximum height 39.4 and minimum height 2.5 on
atomic number scale
 Heavy-Atom Location for CUHABS IN I2/A
   2381 reflections used for structure factor sums
 Solution 1 CFOM = 65.03 PATFOM = 99.9 Corr. Coeff. = 80.7 SYMFOM = 99.9
 Shift to be added to superposition coordinates:  0.3069  0.2500  0.3675
 Name At.No.  x y z   s.o.f.  Minimum distances / PATSMF (self first)
 CU1    31.0  0.3073  0.3269  0.1172  1.0000     3.44
                                                 159.4
```

Copper atom position together with nearest symmetry related atom.

```
 S2     19.0  0.3372  0.4814  0.1487  1.0000     4.52      2.28
                                                 28.4     137.3
```

Sulfur position as for copper and distance from nearest Cu etc.

```
 S3     11.3  0.4617  0.3182  0.1504  1.0000     6.25      1.95  2.80
                                                  0.0      60.8  13.7
```

This must represent O or N in fact, as it is so close to S

```
 O4      8.8  0.1515  0.3218  0.0827  1.0000     3.34      1.97  3.34  3.42
                                                  0.0      69.5  12.0   3.0
 O5      8.2  0.5963  0.2140  0.1818  1.0000     4.31      4.00  4.11  2.27  4.50
                                                       0.0 42.1 16.1 1.9 1.1
 O6      8.0  0.1309  0.5190  0.1171  1.0000     4.37      3.51  2.63  3.17  2.87  3.94
                                                  0.0      36.5  31.5   0.0   1.7   7.8
```

14. *This 6-atom model could be refined directly with SHELXL and the remaining atoms found in difference Fourier syntheses. Here, a second run with SHELXS uses the tangent formula and Fourier recycling to obtain more atomic positions and hence calculate better phases. Input identical to first run except as shown:*

```
TITL CUHABS IN I2/A
........
UNIT   80  88   8  24  24   8
TEXP 200 6       Use tangent formula to expand structure based on 2 heavy atom positions using 200
                 highest E-values. Recommended number is half of those above 1.5, here 437 (see above)
CU1      3  0.30731  0.32691  0.11725 11.00000    0.04    Heavy atom positions as found.
                 Site occupancy fixed at 1.0, temperature factor assigned to be 0.04 Å²
S2       5  0.33721  0.48144  0.14867 11.00000    0.04
O3       4  0.46171  0.31822  0.15036 11.00000    0.04
O4       4  0.15149  0.32178  0.08273 11.00000    0.04
O5       4  0.59628  0.21403  0.18183 11.00000    0.04
O6       4  0.13089  0.51898  0.11706 11.00000    0.04
HKLF   4
```

The output produced is now:

```
TITL CUHABS IN I2/A
CELL  1.54180    12.353    14.315    13.670     90.00    91.62    90.00
........
SUMMARY OF PARAMETERS FOR  CUHA1 IN I2/A
ESEL  Emin  1.200    Emax  5.000     DelU 0.005     renorm 0.700     axis 0
OMIT  s    4.00    2theta(lim)   180.0
TEXP  na    200    nh    6    Ek 1 .500        Note values for na and nh as input above.
FMAP  code  9                                   Program will do three cycles of refinement
PLAN  npeaks    -24     del1 0.500     del2 1.500
MORE  verbosity  1
TIME  t      9999999.
RE = 0.252 for    6 atoms and   513 E greater than   1.500
                              First structure factor calculation — Only input heavy atoms used
Tangent expanded to  758  out of  758  E greater than  1.200
FMAP and GRID set by program
FMAP     9   3  17
GRID     -1.786 -2  -1       1.786   2   1
E-Fourier for CUHABS IN I2/A     First Fourier cycle done, based on tangent recycling
 Maximum =   620.68,   minimum = -132.67
Peak list optimization           Five new peaks accepted as (carbon) atoms.
 RE = 0.194 for   11 surviving atoms and   758 E-values
E-Fourier for CUHABS IN I2/A     Second Fourier calculation
 Maximum =   619.06,   minimum = -141.23
Peak list optimization      Seven more atoms identified in map
 RE = 0.124 for   18 surviving atoms and   758 E-values
E-Fourier for CUHABS IN I2/A  Final Fourier cycle and peaks from map
 Maximum =   621.46,   minimum = -144.91
                   resulting map is almost identical to that produced by the direct methods run.
```

15. *From either the direct methods or the Patterson approach, the output file CUHABS.RES contains the "atoms" that have been found. It can easily be edited manually to become the new CUHABS.INS for the first refinement cycle, using SHELXL [68]. It is only necessary to change the peak numbers Q1, Q2, etc. into atom names O1, N3, etc. and possibly to sort them into a more convenient order. It is also essential to change the number following each atom name to correspond to the number of the atom type as given on the SFAC instruction. By default, these are all set to be 1, normally C.*

```
TITL  CUHABS IN I2/A
CELL  0.71073   12.3543  14.3081  13.5164  90  91.352  90
ZERR  8          0.0012   0.0014   0.0014   0   0.012   0
```

```
LATT  2
SYMM  0.5+X, -Y, Z
SFAC     C   H   N   O   S  CU
UNIT    80  88  24  24   8   8
L.S. 6
FMAP 2
PLAN 20
CU1    6  0.3055  0.3267  0.1193 11.000000  0.05  Atomic coordinates: 6th item is the occupancy
S2     5  0.3381  0.4802  0.1480 11.000000  0.05  factor (10+ value) implies that it is held constant
Q1     1  0.4622  0.3202  0.1483 11.000000  0.05   192.41  7th item is the estimated isotropic atom
Q2     1  0.1562  0.3254  0.0794 11.000000  0.05   188.67  displacement parameter
Q3     1  0.1278  0.5205  0.1189 11.000000  0.05   104.54
Q4     1  0.2219  0.5533  0.1557 11.000000  0.05   100.48  (peak heights from Fourier map —
Q5     1  0.5993  0.2140  0.1817 11.000000  0.05    96.17  they are ignored and may be
Q6     1  0.2385  0.6351  0.1935 11.000000  0.05    87.39  removed)
.....
Q23    1  0.1020  0.5600  0.1783 11.000000  0.05    44.66
Q24    1  0.1864  0.4828  0.1133 11.000000  0.05    42.15
Q25    1 -0.2245  0.1334  0.0800 11.000000  0.05    41.04
HKLF 4
END
```

16. *Refinement with SHELXL. The following abbreviated output from a successful run (CUHABS.LST) of this program begins with a summary of the input data (CUHABS.INS). The model determined by direct methods at 11 has been used here.*

```
 TITL  CUHABS IN I2/A
 CELL  0.71073     12.3543  14.3081  13.5164  90  91.352  90
 ZERR  8             0.0012   0.0014   0.0014   0   0.012   0
 LATT  2
 SYMM  0.5+X, -Y, Z
 SFAC     C   H   N   O   S  CU
 UNIT    80  88  24  24   8   8
V = 2388.58 F(000) = 1288.0 Mu = 2.01 mm-1 Cell Wt = 2534.54 Rho = 1.762
 L.S. 8                                   Number of cycles of least squares
 FMAP 2                                   Calculate difference Fourier Synthesis
 PLAN 20                                  Locate 20 strongest peaks on this map
 FVAR        1.00000
CU1  6  0.30550  0.32670  0.11930  11.00000  0.05
 S2  5  0.33810  0.48020  0.14800  11.00000  0.05
 O2  4  0.46220  0.32020  0.14830  11.00000  0.05
 O1  4  0.15620  0.32540  0.07940  11.00000  0.05
 N2  3  0.12780  0.52050  0.11890  11.00000  0.05
 C10 1  0.22190  0.55330  0.15570  11.00000  0.05
 O3  4  0.59930  0.21400  0.18170  11.00000  0.05
 N3  3  0.23850  0.63510  0.19350  11.00000  0.05
 C9  1  0.50080  0.23630  0.16400  11.00000  0.05
 N1  3  0.31130  0.18670  0.13600  11.00000  0.05
 C1  1  0.09360  0.25970  0.08560  11.00000  0.05
 C7  1  0.23270  0.13750  0.12800  11.00000  0.05
 C8  1  0.41710  0.15390  0.15630  11.00000  0.05
 C5  1 -0.10120  0.20580  0.07410  11.00000  0.05
 C3  1  0.03860  0.09320  0.10860  11.00000  0.05
 C2  1  0.11800  0.16380  0.10990  11.00000  0.05
 C6  1 -0.02480  0.27330  0.07640  11.00000  0.05
 C4  1 -0.07520  0.11540  0.09280  11.00000  0.05
 HKLF 4
Inconsistent equivalents etc.  Note that 1 1 0 has been measured twice with high inconsistency.
   h   k   l      Fo^2   Sigma(Fo^2)  N  Esd of mean(Fo^2)
   1   1   0     244.44      4.42     2      172.11
   1   7   0    6708.11     12.09     5       83.27
   7   7   0    3528.96      8.82     5       67.27
.....
```

The results of the eight refinement cycles follow. At the start, the R-factor and the parameter shifts are both large, but the refinement converges well.

```
Least-squares cycle 1 Maximum vector length = 511 Memory required = 1174 / 102492
 wR2 =  0.5958 before cycle   1 for   2755 data and    73 /    73 parameters
 GooF = S = 6.309; Restrained GooF = 6.309 for 0 restraints
 Weight = 1/[sigma^2(Fo^2)+(0.1000*P )^2+0.00*P] where P = (Max(Fo^2,0)+2*Fc^2 )/3
 Shifts scaled down to reduce maximum shift/esd from   37.76  to    15.00
     N       value        esd    shift/esd  parameter
     1     0.48456     0.00853    -19.905    OSF
     5     0.04002     0.00089    -11.250    U11 Cu1
     9     0.04205     0.00107     -7.439    U11 S2
    13     0.04163     0.00240     -3.482    U11 O2
    17     0.04133     0.00242     -3.581    U11 O1
    25     0.03984     0.00303     -3.356    U11 C10
    41     0.04133     0.00279     -3.103    U11 N1
 Mean shift/esd =    1.543    Maximum = -19.905 for  OSF
 Max. shift = 0.021 A for N1      Max. dU =-0.010 for C10
Least-squares cycle 2
 wR2 =  0.4758 before cycle   2 for   2755 data and    73 /    73 parameters
 GooF = S = 4.182; Restrained GooF = 4.182 for 0 restraints
 Weight = 1/[sigma^2(Fo^2)+(0.1000*P )^2+0.00*P] where P = (Max(Fo^2,0)+2*Fc^2 )/3
 Shifts scaled down to reduce maximum shift/esd from   36.45  to    15.00
     N       value        esd    shift/esd  parameter
     1     0.41248     0.00529    -13.616    OSF
     2     0.30628     0.00011      3.732     x  Cu1
     4     0.11864     0.00010     -3.642     z  Cu1
     5     0.03060     0.00063    -15.000    U11 Cu1
     9     0.03441     0.00079     -9.670    U11 S2
    13     0.03331     0.00179     -4.648    U11 O2
    17     0.03329     0.00181     -4.452    U11 O1
    21     0.03684     0.00210     -3.181    U11 N2
    25     0.03045     0.00222     -4.235    U11 C10
    29     0.03711     0.00177     -3.653    U11 O3
    37     0.03187     0.00230     -3.920    U11 C9
    41     0.03266     0.00208     -4.160    U11 N1
    45     0.03262     0.00230     -3.810    U11 C1
    49     0.03351     0.00232     -3.480    U11 C7
    53     0.03382     0.00237     -3.341    U11 C8
    65     0.03331     0.00244     -3.464    U11 C2
 Mean shift/esd =    1.874    Maximum = -15.000 for  U11 Cu1
 Max. shift = 0.024 A for C1      Max. dU =-0.009 for Cu1
.....
Least-squares cycle   7
 wR2 =  0.2392 before cycle   7 for   2755 data and    73 /    73 parameters
 GooF = S = 1.968; Restrained GooF = 1.968 for 0 restraints
 Weight = 1/[sigma^2(Fo^2)+(0.1000*P )^2+0.00*P] where P = (Max(Fo^2,0)+2*Fc^2 )/3
     N       value        esd    shift/esd  parameter
     1     0.33483     0.00185 -0.005     OSF
 Mean shift/esd =    0.007    Maximum =    0.026 for    z  N3
 Max. shift = 0.000 A for N3      Max. dU = 0.000 for C6
Least-squares cycle   8
 wR2 =  0.2392 before cycle   8 for   2755 data and    73 /    73 parameters
 GooF = S = 1.968; Restrained GooF = 1.968 for 0 restraints
 Weight = 1/[sigma^2(Fo^2)+(0.1000*P )^2+0.00*P] where P = (Max(Fo^2,0)+2*Fc^2 )/3
     N       value        esd    shift/esd  parameter
     1     0.33483     0.00184     0.000     OSF
 Mean shift/esd =    0.001    Maximum =    0.005 for    z  N3
 Max. shift = 0.000 A for N3      Max. dU = 0.000 for C3
 Largest correlation matrix elements
0.885 U11 Cu1 / OSF         0.702 U11 S2 / OSF          0.697 U11 S2 / U11 Cu1
```

Final list of the refined atom parameters with their standard deviations beneath each value. (The first value given is the r.m.s. error for the atomic position in Å.) The displacement parameters, U, are all reasonable and show no great variation, indicating that the atoms are correctly identified.

```
CUHABS IN I2/A
 ATOM              x          y          z           sof          U11     .....
 Cu            0.30688    0.32637    0.11806     1.00000      0.01744
   0.00095     0.00005    0.00004    0.00004     0.00000      0.00027
```

```
S            0.33794   0.48099   0.14862     1.00000     0.02330
  0.00217    0.00011   0.00009   0.00009     0.00000     0.00035
O2           0.46248   0.31556   0.14732     1.00000     0.02000
  0.00617    0.00032   0.00023   0.00027     0.00000     0.00073
O1           0.15631   0.32791   0.08026     1.00000     0.02162
  0.00627    0.00033   0.00023   0.00028     0.00000     0.00078
N2           0.12686   0.52317   0.12358     1.00000     0.02580
  0.00790    0.00039   0.00032   0.00034     0.00000     0.00095
C10          0.22392   0.55007   0.15536     1.00000     0.01816
  0.00795    0.00039   0.00032   0.00035     0.00000     0.00089
O3           0.59441   0.21325   0.18287     1.00000     0.02654
  0.00665    0.00032   0.00027   0.00029     0.00000     0.00083
N3           0.23698   0.63464   0.19492     1.00000     0.02907
  0.00861    0.00042   0.00035   0.00035     0.00000     0.00100
C9           0.49896   0.23248   0.16305     1.00000     0.01896
  0.00817    0.00041   0.00033   0.00035     0.00000     0.00093
N1           0.30944   0.19175   0.13359     1.00000     0.01852
  0.00732    0.00035   0.00031   0.00031     0.00000     0.00082
C1           0.08710   0.25797   0.08783     1.00000     0.01946
  0.00831    0.00042   0.00033   0.00036     0.00000     0.00093
C7           0.22797   0.13550   0.12811     1.00000     0.02126
  0.00853    0.00042   0.00035   0.00036     0.00000     0.00096
C8           0.41681   0.15378   0.15629     1.00000     0.02236
  0.00894    0.00045   0.00036   0.00038     0.00000     0.00100
C5          -0.10112   0.20732   0.07503     1.00000     0.03022
  0.01022    0.00051   0.00041   0.00044     0.00000     0.00118
C3           0.03758   0.09277   0.10957     1.00000     0.02938
  0.00971    0.00048   0.00040   0.00043     0.00000     0.00114
C2           0.11843   0.16310   0.10830     1.00000     0.02042
  0.00852    0.00045   0.00033   0.00037     0.00000     0.00098
C6          -0.02442   0.27643   0.07285     1.00000     0.02512
  0.00920    0.00045   0.00037   0.00040     0.00000     0.00107
C4          -0.07011   0.11311   0.09342     1.00000     0.03089
  0.01009    0.00051   0.00041   0.00044     0.00000     0.00119
Final Structure Factor Calculation for CUHABS IN I2/A
 Total number of l.s.parameters = 73
 wR2 =  0.2392 before cycle   9 for    2755 data and      0 /     73 parameters
 GooF = S = 1.968; Restrained GooF = 1.968 for 0 restraints
 Weight = 1/[sigma^2(Fo^2)+(0.1000*P )^2+0.00*P] where P = (Max(Fo^2,0)+2*Fc^2)/3
 R1 =  0.0856 for   2366 Fo > 4sig(Fo)  and  0.0951 for all   2755 data
 wR2 = 0.2392, GooF = S = 1.968, Restrained GooF = 1.968 for all data
```

Note that the discrepancy indices wR_2 and R_1 are already quite low.

```
Occupancy sum of asymmetric unit = 18.00 for non-hydrogen and 0.00 for hydrogen atoms
 Recommended weighting scheme:  WGHT        0.1822       11.6948
 Note that in most cases convergence will be faster if fixed weights (e.g. the
 default WGHT 0.1) are retained until the refinement is virtually complete, and
 only then should the above recommended values be used.
```

In this case, they may be applied now, as refinement has proceeded very well.

```
 Most Disagreeable Reflections (* if suppressed or used for Rfree)
```

"Disagreeable" is a joke for "discordant" in standard English. A few deviations up to $\Delta F^2/\sigma = 6$ are normal. 110 is probably affected by the beam stop and should be deleted — see above.

```
      h   k   l        Fo^2          Fc^2   Delta(F^2)/esd  Fc/Fc(max)  Resolution(A)
      1   1   0      2180.37      77468.24        7.10        0.966        9.35
      4   4   2      1272.44        466.00        5.13        0.075        2.20
      2   4   2      2004.37        775.80        5.06        0.097        2.80
      4   5   1       459.20       1236.24        3.92        0.122        2.07
     -2   4  14       306.69        885.15        3.67        0.103        0.92
     -4   2  16       593.48       1509.79        3.57        0.135        0.81
      2   4  16      1320.94       2871.05        3.34        0.186        0.81
FMAP and GRID set by program
```

Beginning of a difference Fourier synthesis based on the refined model.

```
 FMAP   2   3  18
 GRID     -1.667 -2  -1      1.667   2   1
 R1 =  0.0943 for    2755 unique reflections after merging for Fourier
 Electron density synthesis with coefficients Fo-Fc
 Highest peak    4.05  at  0.3012  0.3269  0.1525  [  0.47 A from CU1 ]
 Deepest hole   -2.59  at  0.3073  0.2907  0.1170  [  0.51 A from CU1 ]
```

```
Mean =     0.00,    Rms deviation from mean =     0.28 e/A^3
```

Since the highest peaks are now 3-4 e Å^{-3} and are very near to the heavy atoms, they do not indicate missed non-hydrogen atoms. The model is still too crude for the location of H-atoms.

```
Fourier peaks appended to .res file
x        y        z    sof  U   Peak     Distances to nearest atoms
(including symmetry equivalents)
Q1  1  0.3012  0.3270  0.1525  1.0 0.05 4.05 0.47 CU1 1.95 N1 2.00 O2 2.02 O1
Q2  1  0.3112  0.3281  0.0824  1.0 0.05 3.52 0.49 CU1 1.91 O1 2.05 O2 2.07 N1
Q3  1  0.3410  0.4863  0.1138  1.0 0.05 3.40 0.48 S2 1.81 C10 2.33 CU1 2.70 N2
Q4  1  0.3358  0.4776  0.1846  1.0 0.05 3.34 0.49 S2 1.76 C10 2.37 CU1 2.56 N3
Q5  1  0.2412  0.6432  0.1618  1.0 0.05 1.55 0.47 N3 1.35 C10 2.28 N2 2.62 S2
........
```

17. *At the end of the run, the new values of the refined parameters, together with the same general instructions as before, are written to a new file CUHABS.RES, which can then be edited and copied back into CUHABS.INS for the next stage of refinement. In this cycle, the instruction ANIS has been added, which causes the atoms to be refined with anisotropic atomic displacement parameters. The recommended "WGHT" instruction, which the program places at the end of the output has been copied to the beginning of the file in order to use a more sensible weighting scheme. If it has not already been done, this is a good point to sort the atoms into a reasonable order. Inspection of the output CUHABS.LST after this round shows such good R-factors ($wR_2 = 13.1\,\%$, $R_1 = 3.5\,\%$) that it is probable that the hydrogen atoms may be located in the next round of refinement. Adding the instruction 'PLAN 20' will display the 20 strongest peaks in the difference Fourier synthesis calculated after this round. To help with this, we update the weighting scheme. The following extract from the file CUHABS.LST does indeed show the positions of all of the H-atoms with their neighbors. A graphical representation of this map alongside the corresponding F_o synthesis is shown as Fig. 8.1.*

```
.....
 OMIT  1  1  0
 ANIS
 L.S. 6
 FMAP 2
 PLAN 20
 WGHT       0.1822      11.6948
 FVAR        0.33483
 CU1   6    0.306878    0.326369    0.118060    11.00000    0.01744
 S2    5    0.337938    0.480992    0.148622    11.00000    0.02330
.....
HKLF 4
```

Anisotropic displacement parameters now appear

```
   CUHABS IN I2/A
ATOM        x         y         z        sof      U11      U22      U33       U23
 Cu1      0.30684  0.32640  0.11812  1.00000  0.01430 0.01264  0.02581  -0.00047 .....
0.00049    0.00002  0.00002  0.00002  0.00000  0.00021 0.00021  0.00022   0.00010 .....
 S2       0.33794  0.48096  0.14853  1.00000  0.01503 0.01444  0.04227  -0.00417 .....
0.00110    0.00005  0.00004  0.00006  0.00000  0.00032  0.00031 0.00041   0.00025 .....
........
 Final Structure Factor Calculation for CUHABS IN I2/A
 Total number of l.s. parameters =    163  ...
 wR2 =  0.1310 before cycle   7 for   2754 data and     0 /   163 parameters
 GooF = S = 0.572; Restrained GooF = 0.572 for 0 restraints
 Weight = 1/[sigma^2(Fo^2)+(0.1822*P)^2+11.69*P] where P=(Max(Fo^2,0)+2*Fc^2)/3
 R1 =  0.0354 for   2366 Fo > 4sig(Fo)  and  0.0416 for all   2754 data
 wR2 = 0.1310, GooF = S = 0.572, Restrained GooF = 0.572 for all data
........
 Electron density synthesis with coefficients Fo-Fc
 Highest peak    0.82  at  0.0622  0.0266  0.1246  [  1.01 A from C3 ]
```

```
Deepest hole   -0.32  at  0.2036  0.2376  0.1320  [  1.47 A from N1 ]
Mean =   0.00,   Rms deviation from mean =   0.11 e/A^3
Fourier peaks appended to .res file
```

The first 11 peaks are the required H-atoms, two on N2, N3 and C8, one on C3, C4, C5, C6 and C7)

```
              x        y        z  sof  U  Peak     Distances to nearest atoms
                                                (including symmetry equivalents)
Q1  1  0.0622  0.0266  0.1246 1.00 0.05 0.82 >1.01 C3 2.09 C4 2.09 C2  2.57 C7
Q2  1  0.3000  0.6498  0.2186 1.00 0.05 0.80 >0.86 N3 1.90 C10 2.06 O3 2.64 S2
Q3  1  0.4199  0.1236  0.2158 1.00 0.05 0.79 >0.92 C8 1.98 C9 1.99 N1  2.56 O3
Q4  1 -0.1236  0.0672  0.0975 1.00 0.05 0.77 >0.93 C4 2.03 C3 2.04 C5  3.25 C6
Q5  1  0.0729  0.5589  0.1294 1.00 0.05 0.75 >0.84 N2 1.90 C10 2.27 O2 2.44 N3
Q6  1  0.1192  0.4724  0.0946 1.00 0.05 0.75 >0.83 N2 1.88 C10 2.13 O1 2.78 S2
Q7  1  0.4417  0.1137  0.1015 1.00 0.05 0.74 >0.98 C8 2.02 C9 2.03 N1  2.59 O3
Q8  1 -0.1716  0.2225  0.0687 1.00 0.05 0.72 >0.90 C5 1.97 C6 2.03 C4  2.91 N3
Q9  1  0.2360  0.0736  0.1367 1.00 0.05 0.67 >0.90 C7 1.92 N1 1.98 C2  2.48 C3
Q10 1  0.1886  0.6717  0.1942 1.00 0.05 0.66 >0.80 N3 1.87 C10 2.02 O3 2.45 N2
Q11 1 -0.0434  0.3342  0.0611 1.00 0.05 0.62 >0.87 C6 1.97 C1 1.97 C5  2.48 O1
Q12 1  0.1414  0.3299  0.1099 1.00 0.05 0.51  0.45 O1 1.26 C1 2.05 CU1 2.23 C6
Q13 1  0.0787  0.1199  0.1058 1.00 0.05 0.43  0.64 C3 0.79 C2 1.85 C4  1.87 C7
Q14 1  0.2634  0.1639  0.1388 1.00 0.05 0.41  0.62 C7 0.70 N1 1.84 C2  1.91 C8
.....
```

Experimental C—H, N—H and O—H bond lengths are 0.8–1.2 Å

18. *The H-atom positions are now added as instructions to the input data for the next round. An attempt to refine them with individual isotropic atomic displacement factors was not entirely satisfactory, and a large spread of U-values was obtained. Each H-atom was thus constrained to have a displacement parameter 1.2 times the equivalent isotropic value of the atom to which each is bonded, and this is accomplished by giving the value of '−1.2' in place of a U-value. For the final cycles, the evidently badly measured reflection 110 (bad agreement of symmetry equivalents and large $F_o^2 - F_c^2$) is omitted. The weighting scheme is updated and for this last round, the instruction ACTA is included to ensure that all crystallographic data are written to a .cif-file. The instruction BOND $H ensures that all distances to H-atoms will be included. The following (abbreviated) output is produced:*

```
........
OMIT  1  1  0
L.S. 6
BOND  0.5 $H
ACTA
FMAP 2
PLAN 10
WGHT       0.0817      1.8906
FVAR        0.33553
CU    6     0.306841     0.326401     0.118116     11.00000     0.01430     0.01264 =
        0.02581   -0.00047   -0.00124   -0.00035
........
N2    3     0.126588     0.523021     0.123710     11.00000     0.01677     0.02248 =
        0.04041   -0.00938   -0.00536    0.00143
H21   2    0.0729  0.5589  0.1294  11.00000 -1.20     0.75
H22   2    0.1192  0.4724  0.0946  11.00000 -1.20     0.75
N3    3     0.237009     0.634420     0.194607     11.00000     0.02037     0.02029 =
        0.04902   -0.01410   -0.00349    0.00147
H31   2    0.3000  0.6498  0.2186  11.00000 -1.20     0.80
H32   2    0.1886  0.6717  0.1942  11.00000 -1.20     0.66
........
C10   1     0.224390     0.550082     0.155225     11.00000     0.01785     0.01583 =
        0.02314   -0.00024   -0.00092   -0.00020
HKLF 4
........
Least-squares cycle   6      Maximum vector length =
wR2 =  0.0733 before cycle   6 for   2754 data and   196 /   196 parameters
GooF = S = 0.675; Restrained GooF = 0.675 for 0 restraints
```

```
 Weight = 1/[sigma^2(Fo^2)+(0.0817*P )^2+1.89*P] where P = (Max(Fo^2,0)+2*Fc^2)/3
     N      value        esd    shift/esd  parameter
     1     0.33716     0.00057     0.000      OSF
 Mean shift/esd =   0.001    Maximum =  -0.035 for   y  H22
 Max. shift = 0.001 A for H22       Max. dU = 0.000 for N2
 Largest correlation matrix elements
 0.607 U22 Cu / OSF        0.607 U11 Cu / OSF        0.603 U33 Cu / OSF
CUHABS IN I2/A                Definitive list of parameters at end of refinement.
 ATOM x y z sof U11 U22 U33 U23 U13 U12 Ueq
Cu 0.30681 0.32640 0.11812 1.00000 0.01440 0.01273 0.02599
-0.00049 -0.00127 -0.00033 0.01773
0.00032 0.00002 0.00001 0.00002 0.00000 0.00012 0.00012 0.00013
0.00007 0.00009 0.00006 0.00009
.....
N2 0.12733 0.52309 0.12389 1.00000 0.01694 0.02038 0.04427 -0.01102 -0.00515
0.00277 0.02729
0.00264 0.00013 0.00011 0.00014 0.00000 0.00073 0.00071 0.00096 0.00066
0.00069 0.00056 0.00035
 H21          0.08306  0.56008  0.12726    1.00000     0.03274
   0.04261    0.00213  0.00188  0.00183    0.00000     0.00000
 H22          0.12064  0.46926  0.09797    1.00000     0.03274
   0.04533    0.00202  0.00181  0.00188    0.00000     0.00000
........
 Final Structure Factor Calculation for CUHABS IN I2/A
 Total number of l.s. parameters =   196
 wR2 =  0.0733 before cycle   7 for    2754 data and       0 /    196 parameters
 GooF = S = 0.676; Restrained GooF = 0.676 for 0 restraints
 Weight = 1/[sigma^2(Fo^2)+(0.0817*P)^2+1.89*P] where P=(Max(Fo^2,0)+2*Fc^2)/3
 R1 =  0.0244 for   2366 Fo > 4sig(Fo)  and  0.0302 for all   2754 data
 wR2 = 0.0733, GooF = S = 0.676, Restrained GooF = 0.676 for all data
                     R-values, goodness-of-fit and data : parameter ratio all satisfactory
......
 Principal mean square atomic displacements U
   0.0263   0.0143   0.0126   Cu  principal components of vibration ellipsoids physically reasonable
   0.0434   0.0149   0.0139   S
   0.0380   0.0175   0.0140   O1
   0.0317   0.0165   0.0152   O2
........
Most Disagreeable Reflections (* if suppressed or used for Rfree)  No "outriders"!
     h   k   l           Fo^2          Fc^2   Delta(F^2)/esd    Fc/Fc(max)    Resolution(A)
     3   5   2        2024.24        1469.86        5.67          0.140         2.21
     4   5   1         452.86         698.27        5.60          0.096         2.07
    13   9   6         981.37        1328.88        4.30          0.133         0.76
     1   8   1         630.26         467.70        4.27          0.079         1.75
........
Bond lengths and angles          Values and standard deviations all satisfactory
 Cu -        Distance        Angles
 O1        1.9169 (0.0013)
 N1        1.9450 (0.0015)    93.19 (0.05)
 O2        1.9642 (0.0013)   174.63 (0.05)   83.46 (0.05)
 S         2.2800 (0.0005)   101.21 (0.04)  160.32 (0.05)   83.05 (0.04)
               Cu -          O1             N1              O2
........
R1 =  0.0293 for   2754 unique reflections after merging for Fourier
Electron density synthesis with coefficients Fo-Fc
 Highest peak     0.40  at  0.0808  0.1258  0.1014  [  0.71 A from C2 ]
 Deepest hole    -0.28  at  0.2007  0.2384  0.1428  [  1.51 A from N1 ]
 Mean =     0.00,   Rms deviation from mean =      0.07 e/A^3
                              Maximum residual electron density 0.40 e Å^-3 .
        x       y       z    sof    U   Peak   Distances to nearest atoms
                                                (including symmetry equivalents)
Q1 1 0.0808 0.1258 0.1014 1.00 0.05 0.40 0.71 C2 0.72 C3 1.43 H3 1.85 C7
Q2 1 0.1456 0.3287 0.1120 1.00 0.05 0.39 0.45 O1 1.28 C1 1.99 CU 2.04 H22
Q3 1 0.3171 0.3259 0.1783 1.00 0.05 0.38 0.82 CU 1.86 O2 2.02 N1 2.27 S
.......
```

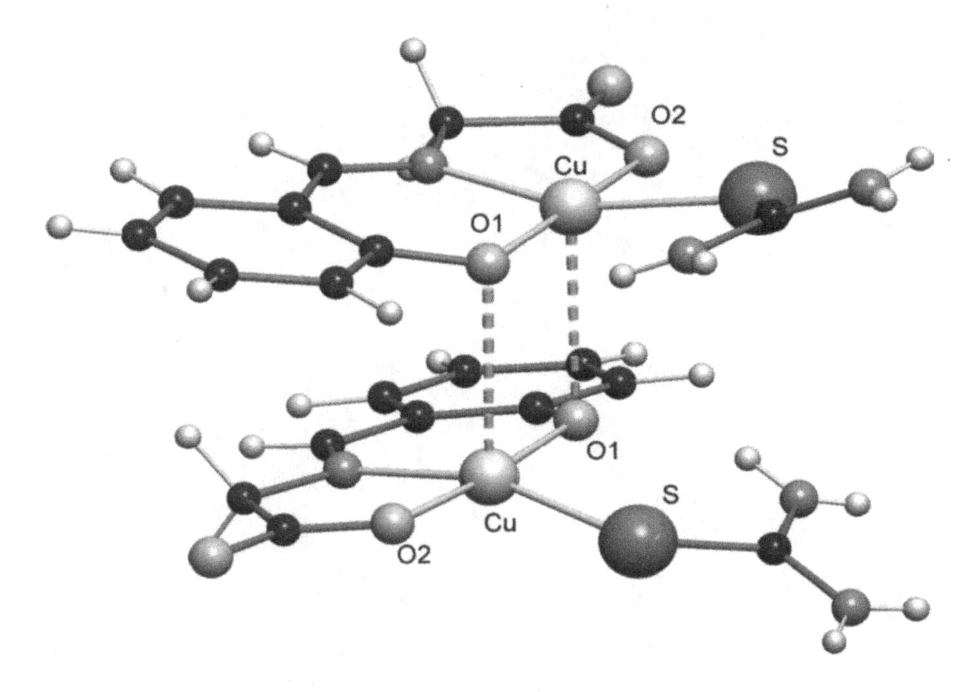

Fig. 15.3. Ball and spoke model of a CUHABS-dimer (DIAMOND [78]).

19. *The formal structure determination is now complete, and data have been produced for structural interpretation and preparation of diagrams. Examples of various possibilities of these are given in chapter 12. They show, among other things, how important the consideration of the crystal symmetry is, which here leads to the formation of discrete dimers by action of the 2-fold axis.*

Bibliography

Selection of crystallographic texts

[1] W. Borchardt-Ott, *Crystallography,* Springer-Verlag, 2nd Edition, 1995.
[2] W. Clegg, *Crystal Structure Determination,* Oxford University Press, 1998.
[2a] W. Clegg, A. J. Blake, R. O. Gould, P. Maine, *Crystal Structure Analysis: Principles and Practice,* Oxford Univ. Press, 2001.
[3] J. D. Dunitz, *X-Ray Analysis and Structure of Organic Molecules,* Wiley-VCH 1995, 2nd Edition, 1995.
[4] C. Giacovazzo, Ed., *Fundamentals of Crystallography,* Oxford University Press, 2nd Ed. 2002.
[5] J. P. Glusker and K. N. Trueblood, *Crystal Structure Analysis,* Oxford University Press, 1985.
[6] J. P. Glusker, M. Lewis, M. Rossi, *Crystal Structure Analysis for Chemists and Biologists,* Verlag Chemie, 1994.
[7] M. F. C. Ladd, *Symmetry in Molecules and Crystals,* Ellis Horwood, 1989.
[8] M. F. C. Ladd and R. A. Palmer, *Structure Determination by X-Ray Crystallography,* Kluwer Academic, 4th Ed. 2003.
[9] P. Luger, *Modern X-Ray Analysis on Single Crystals,* W. de Gruyter, Berlin 1980.
[10] G. H. Stout, L. H. Jensen, *X-Ray Structure Determination.* 2nd Edition, Wiley & Sons, New York 1989.
[11] M. Woolfson, *X-Ray Crystallography,* Cambridge, 1997.

Tables and Data Collections

[12] International Tables of Crystallography,
Vol. A: Space Group Symmetry, T. Hahn Ed., 5th Ed., Kluwer 2002.
Vol. B: Reciprocal Space, U. Shmueli Ed., 2nd Ed., Kluwer 2001.
Vol. C: Mathematical, Physical and Chemical Tables, E. Prince & A. C. Wilson Eds., 2nd Ed., Kluwer 1999.
Vol. D: Physical Properties of Crystals, A. Authier, Ed., Kluwer 2003.
Vol. E: Subperiodic Groups, V. Kopsky & D. B. Litvin, Eds., Kluwer 2002.
Vol. F: Crystallography of Biological Macromolecules, M. G. Rossmann & E. Arnold Eds., Kluwer 2001.
Forthcoming Volumes:
Vol. A1: Symmetry Relations between Space Groups, ed. by H. Wondratschek & U. Müller
Vol. G: Definition and Exchange of Crystallographic Data, ed. by S. R. Hall & B. McMahon
[13] Strukturbericht **1–7**, Akad. Verlagsges., Leipzig 1931–1943; then Structure Reports from **8**, Kluwer, Dordrecht from 1956.

[14] O. Kennard et al., Eds., Molecular Structures and Dimensions, D. Reidel, Dordrecht from 1970.
[15] R. W. G. Wyckoff, Crystal Structures, Vol 1–6. Wiley & Sons, New York 1962–1971.
[16] Landolt-Börnstein, Zahlenwerte aus Naturwissenschaft und Technik, Neue Serie, III, **Vol. 7**. Springer-Verlag, Berlin 1973–1978.
[17] J. Donohue, The Structures of the Elements, Wiley & Sons, New York 1974.

Other References

[18] P. M. De Wolff et al., Acta Crystallogr. **A48** (1992) 727
[19] C. J. E. Kempster, H. Lipson, Acta Crystallogr. **B28** (1972) 3674
P. Roman, C. Guzman-Miralles, A. Luque, Acta Crystallogr. **B49** (1993) 383
[20] M. Molinier, W. Massa, J. Fluor. Chem., **57** (1992) 139
[21] H. Bärnighausen, Group-Subgroup Relations between Space Groups: a useful tool in Crystal Chemistry. MATCH, Commun. Math. Chem. **9** (1980) 139.
[22] U. Müller, *Inorganic Structural Chemistry*, Wiley, Colchester, 1996.
[23] J. Hulliger, Angew. Chem. **106** (1994) 151
[24] a) R. A. Young, *The Rietveld Method*, Oxford University Press 1993.
b) R. Allmann, Röntgenpulverdiffraktometrie, Springer 2002.
[25] H. Hope, Acta Crystallogr. **A27** (1971) 392
[26] N. Walker, D. Stuart, Acta Crystallogr. **A39** (1983) 158.
[27] A. Mosset, J. Galy, X-Ray Synchrotron Radiation and Inorganic Structural Chemistry. Topics in Current Chemistry, **145** (1988) 1.
[28] G. E. Bacon, *Neutron Scattering in Chemistry*. Butterworths, London 1977
[29] W. Hoppe, Angew. Chem. **59** (1983) 465
[30] D. L. Dorset, S. Hovmöller and X. D. Zou (eds.), Electron Crystallography, Kluwer Academic Publishers, Dordrecht 1997.
[31] J. Karle, H. Hauptman, Acta Crystallogr. **3** (1950) 181.
[32] W. Cochran, M. M. Woolfson, Acta Crystallogr. **8** (1955) 1.
[33] W. H. Zachariasen, Acta Crystallogr. **5** (1952) 68.
[34] W. H. Baur, D. Kassner, Acta Crystallogr. **B48** (1992) 356.
[35] W. J. Peterse, J. H. Palm, Acta Crystallogr. **20** (1966) 147.
[36] P G. Jones, Acta Crystallogr. **A42** (1986) 57.
[37] A. M. Glazer, K. Stadnicka, Acta Crystallogr. **A45** (1989) 234.
[38] W. C. Hamilton, Acta Crystallogr. **18** (1965) 502.
[39] H. D. Flack, Acta Crystallogr. **A39** (1983) 876.
[40] W. H. Zachariasen, Acta Crystallogr. **23** (1967) 558.
[41] P. Becker, P. Coppens, Acta Crystallogr. **A30** (1974) 129 and 148.
[42] P. Becker, Acta Crystallogr. **A33** (1977) 243.
[43] M. Renninger, Z. Physik **106** (1937) 141.
[44] Y. Laligant, Y. Calage, G. Heger, J. Pannetier, G. Ferey, J. Solid State Chem. **78** (1989) 66.
[44a] R. E. Schmidt, W. Massa, D. Babel, Z. anorg. allg. Chem. **615** (1992) 11.
[45] K. Kirschbaum, A. Martin, A. A. Pinkerton, J. Appl. Cryst. 30 (1997) 514

[46] B. T. M. Willis in: Intern. Tables of Crystallography **Vol. C**, Sect. 7.4.2.
[47] H.-G. v. Schnering, Dong Vu, Angew. Chem. **95** (1983) 421.
[48] D. Babel, Z. anorg. allg. Chem. **387** (1972) 161.
[49] U. Müller, Angew. Chem. **93** (1981) 697.
[50] W. Massa, S. Wocadlo, S. Lotz and K. Dehnicke, Z. anorg. allg. Chem., **589** (1990) 79.
[51] M. Molinier and W. Massa, Z. Naturforsch., **47b**, 783-788 (1992).
[52] W. C. Hamilton, Acta Crystallogr. **14** (1961) 185.
[53] W. C. Hamilton, Acta Crystallogr. **12** (1959) 609.
[54] J. Drenth, Principles of Protein X-ray Crystallography, 2nd ed., Springer, New York 1999.
[55] D. E. McRee, Practical Protein Crystallography, 2nd ed., Academic Press, San Diego 1999.

Selection of Important Programs

(The assistance of Mr L. M. D. Cranswick of CCP14 (Collaborative Computational Project Number 14 for Single Crystal and Powder Diffraction) in compiling this list is gratefully acknowledged. CCP14 is a useful source of information on available computer programs, and may be contacted by Email at: ccp14@ccp14.ac.uk or on the Web at: http://www.ccp14.ac.uk/. A source of Windows applications of many of the programs listed below is Dr L. Farrugia, E-mail: louis@chem.gla.ac.uk, Website: http://www.chem.gla.ac.uk/~louis/software/. Only a single author is given in these references to facilitate contact. Most of the programs are under continuous development.)

[56] LEPAGE: Y. Le Page, Program for Lattice Symmetry Determination, National Research Council of Canada, Ottawa, CANADA K1A 0R6, E-mail: yvon.le_page@nrc.ca. See also [70].
[57] CRYSTALS: D. J. Watkin, General Crystallographic Software, Including Graphics. Chemical Crystallography Laboratory, 9 Parks Road, Oxford OX1 3PD, U. K. E-mail: david.watkin@chem.ox.ac.uk, Website: http://www.xtl.ox.ac.uk/
[58] ORTEPIII: C. K. Johnson, Fortran Thermal Ellipsoid Plot Program. Oak Ridge National Laboratory, Oak Ridge, TN, 37831-6197, U. S. A., E-Mail: ortep@ornl.gov, Website: http://www.ornl.gov/ortep/ortep.html
[59] MISSYM: Y. Le Page, A Computer Program for Recognizing and Correcting Space-Group Errors, see ref. [56]. Extended form included in [70].
[60] DIFABS: N. Walker, Program for Automatic Absorption Correction, see [26]. Included with modifications in [57], [66], [70], and [81].
[61] PATSEE: E. Egert, *Adjunct to [64] for Patterson Search Methods*, Institute of Organic Chemistry, Marie Curie-Str. 11, D-60439 Frankfurt am Main, Germany E-mail bolte@indyl.org.chemie.uni-frankfurt.de, Website: http://www.org.chemie.uni-frankfurt.de/egert/html/patsee.html.
[62] SIMPEL: H. Schenk, Program for Structure Solution using Higher Invariants, University of Amsterdam, Nieuwe Achtergracht 166, 1018 WV Amsterdam, Netherlands, E-Mail: hs@crys.chem.uva.nl.
[63] MULTAN: Program for the Determination of Crystal Structures, Department of Physics, University of York, York YO1 5DD, U. K. E-mail: pml@vaxa.york.ac.uk (This program has spawned a number of variants!)

[64] SHELXS: G. M. Sheldrick, Program for the Solution of Crystal Structures, Institut für Anorganische Chemie der Universität Göttingen, Tammannstr. 4, D-37077 Göttingen, Germany E-mail: gsheldr@shelx.uni-ac.gwdg.de, Website: http://shelx.uni-ac.gwdg.de/SHELX/

[65] SIR: C. Giacovazzo: Integrated program for Crystal Structure solution, Dipartimento Geomineralogico, Campus Universitario, Via Orabona 4, 70125 Bari, Italy, E-Mail: sirmail@area.ba.cnr.it, Website: http://www.irmec.ba.cnr.it

[66] XTAL: S. R. Hall. The X-tal System, Crystallography Centre, The University of Western Australia, Nedlands, 6907, Perth, Australia, E-mail: xtal@crystal.uwa.edu.au, Website: http://xtal.sourceforge.net/

[67] DIRDIF: P. T. Beurskens, Structure Solution Using Difference Structure Factors, Lab. voor Kristallografie, U. Nijmegen, Toernooiveld 6525 ED Nijmegen, Netherlands, E-mail: ptb@sci.kun.nl, Website: http://www-xtal.sci.kun.nl/documents/software/dirdif.html

[68] SHELXL: G. M. Sheldrick, Program for the Refinement of Crystal Structures, (Address, etc., under [64].)

[69] CRUNCH: R. A. G. de Graaff, Integrated Direct Methods Program, Gorlaeus Lab., U. Leiden, Einsteinweg 55, 2300 RA Leiden, Netherlands, E-mail: rag@chemb0b.leidenuniv.nl, Website: http://www.bfsc.leidenuniv.nl/software/crunch

[70] PLATON: A. L. Spek, Program for the Geometric Interpretation of Structural Data, Lab. voor Kristal- en Structuurchemie, U. Utrecht, Padualaan 8, 3584 CH Utrecht , Netherlands, E-mail a.l.spek@chem.uu.nl, Website: http://www.cryst.chem.uu.nl/platon/

[71] SDP, MolEN: Structure Determination Systems for Enraf-Nonius Diffractometers, Website: http://bruker-axs.com

[72] PLUTO: W. D. S. Motherwell, Molecular Plotting Program. Now incorporated in [70].

[73] SCHAKAL97: E. Keller, Computer Program for the Graphic Representation of Molecular and Crystallographic Models, Kristallographisches Institut der Universitat Freiburg, Hebelstr. 25, D-79104 Freiburg im Breisgau, Germany. E-mail: kell@uni-freiburg.de, Website: http://www.krist.uni-freiburg.de/ki/Mitarbeiter/Keller/

[74] SHELXTL Structure Determination systems for Bruker Diffractometers, Website: http://www.bruker-axs.com

[75] XFPA: F. Pavelcik, General Patterson Approach to Structure Solution, Comenius University, Bratislava, Slovak Republic, E-mail: pavelcik@fns.uniba.sk, Website: http://www.fus.uniba.sk/fus/struc-fakag/xfpa.ht

[76] There is an overview of molecular modelling and graphics programs in International Tables B, [12] Section 3.3.

[77] DIRAX: A. J. M. van Duisenberg. A program for indexing twinned crystals, Lab. voor Kristal- en Structuurchemie, U. Utrecht, Padualaan 8, 3584 CH Utrecht , Netherlands, E-mail: duisenberg@chem.uu.nl.

[78] DIAMOND, Visual Crystal Structure Information System, Crystal Impact GbR, Immenburgstr. 20, D-53121 Bonn. E-mail: products@crystalimpact.de, website: www.crystalimpact.com.

[79] JANA, The Crystallographic Computing System: V. Petříček, M. Dušek, Inst. of Physics, Academy of Sciences of the Czech Republic, Cukrovarnika 10, 16253 Praha. E-mail: petricek@fzu.cz, Website: www-xray.fzu.cz/jana/jana.html.

[80] ATOMS: E.Dowty, Shape Software, 521 Hidden Valley Road Kingsport, TN 37663 USA. E-mail: dowty@shapesoftware.com, website: www.shapesoftware.com.

[81] WINGX, A Integrated System of Windows Programs for the Solution, Refinement, and Analysis of Single Crystal Diffraction Data: L. J. Farrugia, Dept. of Chemistry, University of Glasgow, U. K. E-mail: louis@chem.gla.ac.uk, website: www.chem.gla.ac.uk/~louis/wingx

[82] TWINXL, Program for Handling Data of Twinned Crystals: F. Hahn, W. Massa, Fachbereich Chemie, Philipps-Universität, D-35032 Marburg. E-mail: massa@chemie.uni-marburg.de, website: www.chemie.uni-marburg.de/~massa.

[83] SnB, A Direct Methods Procedure for Determining Crystal Structures: C. M. Weeks et al., Hauptman-Woodward Medical Research Institute, Inc., 73 High Street, Buffalo, NY 14203-1196, USA. E-mail: snb-requests@hwi.buffalo.edu, website: www.hwi.buffalo.edu/SnB/Contact.htm.

[84] RASMOL, Molecular Visualization Freeware: www.umass.edu/microbio/rasmol/

Index